IET CIRCUITS, DEVICES AND SYSTEMS SERIES 22

Nanotechnologies

Other volumes in this series:

Nanotechnologies

Michel Wautelet *et al.*

Preface by Jean-Marie Lehn
Member of the Academy of Science
Professor at the Collège de France
Nobel Prize in Chemistry 1987

Translated from French by
Dr Aboubacar Chaehoi

The Institution of Engineering and Technology

Published by The Institution of Engineering and Technology, London, United Kingdom

Second edition © 2006 Dunod, Paris
English translation © 2009 The Institution of Engineering and Technology

Second edition 2006
English translation 2009
Reprinted 2012

The Institution of Engineering and Technology
Michael Faraday House
Six Hills Way, Stevenage
Herts, SG1 2AY, United Kingdom

www.theiet.org

British Library Cataloguing in Publication Data
A catalogue record for this product is available from the British Library

ISBN 978-0-86341-941-6

Typeset in India by Newgen Imaging Systems (P) Ltd, Chennai
First printed in the UK by Athenaeum Press Ltd, Gateshead, Tyne & Wear
Reprinted in India by Replika Press Pvt. Ltd.

Contents

Authors

Dr David Beljonne, Dr Jérôme Cornil, Prof. Philippe Dubois, Damien Duvivier, Prof. Pierre Gillis, Dr Yves Gossuin, Prof. Michel Hecq, Dr Roberto Lazzaroni, Prof. Robert Muller, Dr Assia Ouakssim, Johnny Robert, Dr Alain Roch, Prof. Michel Wautelet work at the University of Mons-Hainaut (Belgium).

Dr Michael Alexandre, Dr Rachel Gouttebaron, Dr Philippe Leclère, Dr Fabien Monteverde work at Matteria Nova, Mons (Belgium).

Prof. Jean-Luc Brédas is Emeritus Professor at the University of Mons-Hainaut (Belgium) and Professor at the University of Arizona.

Chapters 1, 2, 3
Prof. Michel Wautelet

Chapters 4, 5, 6, 7
Dr David Beljonne, Prof. Jean-Luc Brédas, Dr Jérôme Cornil, Dr Roberto Lazzaroni, Dr Philippe Leclère

Chapter 8
Dr Michael Alexandre, Prof. Philippe Dubois

Chapter 9, Appendix 3
Prof. Pierre Gillis, Dr Yves Gossuin, Prof. Robert Muller, Dr Assia Ouakssim, Dr Alain Roch

Chapter 10
Damien Duvivier, Johnny Robert, Prof. Michel Wautelet

Appendix 1
Dr Fabien Monteverde

Appendix 2
Dr Rachel Gouttebaron, Prof. Michel Hecq

Preface to the first edition

Nanosciences and nanotechnologies create research fields of a fundamental interest that could result in the development of many new applications. They also represent subjects that mix together different sciences, including physics, chemistry and biology. For example, they raise challenges in structure, function and implementation domains where chemistry plays the main part in the solution.

In the development process of nanotechnologies, after defining the structure that fulfils the given function arises the issue of manufacturing.

Supramolecular chemistry offers a novel approach requiring auto-arrangement. It is based on the implementation of programmed chemical systems that allow the spontaneous generation of functional supramolecular architectures controlled by the molecular data stored in the components. Therefore, it becomes theoretically possible to avoid resorting to very heavy and expensive processes of nanofabrication and nanomanipulation by letting a system develop by itself, using its properties of auto-fabrication from molecular units that contain the required instructions.

We are grateful to the authors of this book for setting out the basis of this domain. It is on this basis that future devices based on auto-arrangement may be built. This property allows us to go beyond the issues of manufacturing the smallest devices and the most accurate addressing in order to reach the highest level of function and behaviour. Beyond miniaturization comes a new age of increasing complexity.

Jean-Marie Lehn
Nobel Prize in Chemistry 1987
Professor, Collège de France and Louis Pasteur University (Strasbourg)

Strasbourg, 10 January 2003

Introduction

One of the major trends of current science and technology is the movement towards miniaturization. To make smaller and smaller is the key issue for numbers of scientists and industrials; to better understand and use the fundamental laws of the material's behaviour; to make a device faster or less expensive...

In the technology field, making things smaller means working with the atoms, using and handling them. Indeed, it has been some time since physicists, chemists, biologists and engineers have used the atom's properties. However, it is only since 1980 that scientists have been able to deal with atoms one by one, which has in turn led to the fundamental and applied work that could be used in so many branches of industry.

According to the Royal Society of Great Britain, 'nanoscience' is the study of phenomena and the manipulation of material at the atomic, molecular, macromolecular scale, where the properties differ notably from the properties at a bigger scale. The 'nanotechnologies' are the design, characterization, production and implementation of structures, mechanisms and systems by monitoring shape and size at the nanometric scale. To work at these scales is no longer isolated work performed by physicists, chemists, biologists or engineers. Now, all of these disciplines must work together. In nanotechnologies, physical and chemical properties are not separated: they depend on the way the considered systems are synthesized and manufactured.

Nanotechnologies are of particular interest to scientists and engineers. Today, manufacturers are interested, as well as economists, political decision makers and the general public. The topic is so new, its potential applications so varied and its progress so rapid that we feel it is necessary to know more about it.

Nanotechnologies are inextricably a pluridisciplinary field; one can get an exact idea of it only by taking into account all the aspects of each involved field. Like other domains, it is based on general knowledge gathered by scientists and engineers.

Because this branch of industry is so novel and important, it is essential to motivate students and teachers at this early stage. The book's authors teach future physicists, chemists, engineers, biologists and medical practitioners; therefore, they understand the prerequisites very well. Experts in specific fields of nanotechnologies, they teach the basics and demonstrate the present state of research and development. This is the purpose of this book.

The field of nanotechnologies is huge, with many different characteristics. Some of them, such as carbon nanotubes or molecular computers, are already familiar to students, the general public, and policy decision makers through popular scientific papers. However, since it is almost impossible to have a fully comprehensive knowledge of this field, this book deals with other, less common, areas. While economic and cultural applications are important, these lesser-known applications still affect us, and will be studied in depth by future researchers and engineers.

The first chapter, 'The nanotechnology revolution', begins with a historical examination of work on nanotechnologies. At the nanotechnologies scale, some phenomena that are important at our scale become negligible while others become preponderant, with the result being that our experience-based perceptions are no longer valid. In order to understand these differences, the scale rules of various classical physical laws are discussed. The possible application fields of nanotechnologies are briefly presented in order to understand, among other contexts, the economic situation of nanotechnologies.

The second chapter, 'Atomic structure and cohesion', deals with the arrangement of atoms in nanoparticles; the number of atoms at the surface being significant compared with the overall number. Consequently various parameters, such as those relating to consistency, can be different from what they are in a solid. After a summary of the surface properties, the adhesion phenomena are presented. When we examine the nanoparticles, two complementary ways are open for researchers. In the *top-down* approach, we start from the solid state and we study how the properties of a nanoparticle change with the size and the shape. In the *bottom-up* approach, we begin from the atom and we observe the formation of heaps and specific structures (fullerenes, carbon nanotubes, etc.).

The electronic structure of nanoparticles is different from the solids, and is an element that is behind a lot of applications. This is the subject matter of the third chapter, titled 'Electronic structures of nanosystems'. After a summary of essential elementary knowledge of quantum mechanics (especially the behaviour of electrons in matter), we will discuss the effect of size.

Electronics is one of the most active fields in nanotechnologies; this will be discussed in the fourth chapter: 'Molecular electronics'.

One of the characteristics of nanotechnologies is that skills of scientists from many different fields are required. When electronics joins biology, amazing fields are created such as *neuroelectronics*, which is the subject of the fifth chapter.

Until now, most electronics and optics applications relied mostly on inorganic material; however, plastic technology could soon change this, so it is important to understand what it is all about, through its achievements and possibilities. This subject will be discussed in the chapter entitled 'Plastic electronics'.

Thanks to different techniques of *fabrication of nanostructures*, nanotechnologies exist and are expanding today. Those techniques are presented in Chapter 7.

Besides their electronic properties, nanoparticles are also used for their other different properties, such as mechanical properties, in composite materials (the mixture of immiscible materials that gives to the resultant material properties that none of the individual materials has). Among them, nanocomposites with organic matrices are an interesting group; they are discussed in Chapter 8: 'Organic-matrix-based nanocomposites'. After an overview of the different preparation methods, the mechanical properties of nanocomposites as a fluid barrier and their properties of resistance to heat are studied. This demonstrates the diversity of nanotechnologies applications.

Nanoparticles of magnetic materials have properties that are of interest to fields such as biomedicine. This is the subject of Chapter 9, 'Nanomagnetism'. After an overview of the different types of magnetism, the magnetic colloids are presented. The use of nanomagnets in thermotherapy is then presented. Magnetic nanoparticles exist in biology as well. Various cases of biomagnetism are finally described.

Chapter 1
The nanotechnology revolution

'Imagine a machine so small that it is invisible to the naked eye. Imagine those machines with gears not much bigger than a grain of pollen. Imagine those machines manufactured by thousands at the same time at a cost of a few tenths of euros each. Imagine a world so small that gravity has no importance but where atomic forces prevail …'

Paul McWhorter

Just before the year 2000, the kind of machine described above belonged to science fiction. In the movie *Fantastic Voyage*, Richard Fleischer conceived to send a group of five miniaturized surgeons and their submarine into the body of a Czech scientist to destroy a blood clot in his brain. Other authors imagined the use of miniaturized machines to unblock veins clogged with cholesterol with a miniature 'bulldozer', quietly spy on an enemy using a plane as tiny as a fly, or inject a lethal poison through the sting of an artificial wasp.

At the same time, some scientists dreamt of manipulating atoms. In 1959, Richard Feynman, future Nobel Prize winner, gave a talk at the meeting of the American Physical Society entitled 'There is plenty of room at the bottom'. In his talk he proposed that many different end applications could come from the ability to manipulate matter at the nanometric scale. As a theoretician, Feynman did not know how to achieve this; however, he correctly anticipated many possible applications. His talk was given at a time when current processes for fabricating microelectronic chips had not been invented; however, it did not prevent his vision of the future. He concluded that, if we could manipulate atoms, we could write the equivalent of the *Encyclopaedia Britannica* on the head of a pin. Chemistry would also be a matter of single-atom manipulation rather than complex chemical reactions. We could build wires and tools at the microscopic scale, atom by atom.

Fiction started to become reality in the 1980s. In 1981 the Scanning Tunnelling Microscope (STM) was created, earning its inventors G.K Binnig and H. Rohrer from IBM Zürich the 1986 Nobel Prize in Physics. STM comprises a tiny metallic tip so fine that it has only a few atoms, and moves at a fraction of a nanometre from the surface of the object to scan. Thanks to a quantum effect called the *tunnel effect*, electrons can migrate from the tip to the surface. By controlling the movement of the tip over the surface, we can see atoms at their own scale: around a tenth of a nanometre.

A further progression took place in 1990, when D. Eigler and E. Schweizer succeeded in moving xenon atoms one by one, arranging them on a surface of nickel and drawing the initials 'IBM'. By doing this, it became possible to manipulate matter atom by atom, which was the starting point of a new field: nanotechnology.

In the same period, research in nanotechnologies was aided by the discovery of new materials. In 1986, researchers at the Universities of Tucson and Heidelberg discovered molecules composed of 60 carbon atoms in soot, formed in an electric arc between carbon electrodes. They deduced that the C_{60} molecule had a soccer-ball structure, where each atom was located at the junction of three seams, at the top of pentagons and hexagons. In 1996, this discovery earned R. Smalley, H. Kroto and R. Curl the Nobel Prize in Chemistry. Afterwards, many studies demonstrated the surprising physical and chemical properties of these molecules and some other similar ones, known as *fullerenes* (see Chapter 2).

Later, in 1991 in Japan, S. Iijima discovered, in the same soot, tubules later named *carbon nanotubes*. Their structure was that of a rolled-up graphite layer. These examples of tubes measured only 1.5 nm in diameter but could stretch to a few micrometers long (see Chapter 2). Furthermore, they had amazing electrical properties: depending on how the layers were rolled up, the nanotubes could be metallic or semiconducting. Their mechanical properties were also amazing – the mechanical resistance was higher than the mechanical resistance of a steel wire.

Those discoveries led, naturally, to a great deal of prognostication. Even if many of them have been exaggerated, they have contributed ways of demonstrating the existence of new materials, and have made the general public and decision makers aware of a new field of great interest: nanotechnologies.

1.1   From micro- to nanoelectronics

While scientists concentrated their interest on atoms, a new technology field was booming: microelectronics. In order to improve computer features, we need to design increasingly thinner structures. One of the techniques used to achieve this is photolithography. In order to draw a pattern at the micrometer scale several times on the same silicon wafer, a repetitive mask is first created at the macroscopic scale. The mask is then reduced to a small-scale reproduction using photography techniques. This small-scale reproduction mask is placed below a beam of light that is focused on the wafer by means of a dedicated optic system. All of this requires very sharp optic elements (lenses, frame, apertures, etc.) designed with this particular application in mind. Where the wafer is exposed to the beam of light, various physicochemical phenomena occur, which lead to the production of a structure with the required properties.

The physical limit to the size of the structures is fixed by elementary laws of optics. Light is composed of electromagnetic waves. When passing through an aperture or a lens system, light is subjected to diffraction. The ideal outcome is spots of light of a size comparable to the wavelength of the light. The wavelength of visible light is between 0.4 and 0.8 μm. Using this method we cannot draw patterns smaller than 0.5 μm. To achieve smaller structures, ultraviolet rays must be used (or other methods). However, this raises issues of using new techniques. Nevertheless, the field of microelectronics needs to keep moving forward.

Table 1.1   Progress of features of microelectronic devices

Year	1995	1998	2001	2007	2010
Minimum dimension (nm)	150	130	100	70	50
Number of transistors per chip (millions)	40	76	200	520	1400
Frequency (GHz)	1.5	2.1	3.5	6.0	10.0

The evolution in microelectronics follows a trend known as *Moore's law.* In 1965, while working on a presentation about the predictable evolution of electronics, Gordon Moore made an important observation: he noticed, from the growth in the performance of an electronic chip, that every new chip was roughly twice as powerful as the previous one and its development time varied between 18 and 24 months. He observed that, if that trend continued, the performance of computers would exponentially increase over quite short periods. Since this discovery, Moore's law has never been contradicted. In 30 years, the number of integrated transistors in a chip increased from 2 250 (processor 4004, 1971) to 125 million (Pentium® Prescott, 2004). People believe that this trend will go on until 2015. The evolution's forecasts until 2010 are summarized in Table 1.1.

Around 2015, it is likely that we will come up against a brick wall imposed by the laws of physics. What should we do next? Two complementary ways are open for industrialists.

The first solution is to diversify end applications. Microelectronics research has developed techniques that allow devices to be manufactured with micrometre dimensions. Consequently, why not use these microelectronics tools and processes to achieve other purposes? Additionally, this would allow existing, very expensive, high-performance equipment to be profitable much more cost-effectively. Some people have considered using the same techniques to manufacture mobile devices. In 1994, in the United States, John Hunn and his colleagues produced diamond micro-gears. This was the starting point for a new activity field: micro-machines or MEMS (micro-electro-mechanical systems). These are miniature devices with very small components (a few micrometres), and manufactured using the techniques and processes of microelectronics (photolithography etc.). These systems have microscopic moving mechanical parts and integrated electronic circuits, often on the same silicon chip. They often act as sensors for measuring different phenomena (force, movement, illumination, chemical compounds, etc.). Their electronic circuits give them a 'decision-making capability', which allows them to react to events. These devices are genuine 'smart' systems. The MEMS market is booming with end applications in areas as diverse as automobile, space, telecommunications and biomedical industries.

The second solution involves developing the nanotechnology field and achieves Richard Feynman's dream: to devise systems based on nanometre dimension components. Nanosciences and nanotechnologies can be defined as sciences of nanoscopic systems. The prefix 'nano' refers to a nanometre (nm), one thousandth of a micrometre. The systems we are talking about have at least one of their dimensions smaller than a few hundred nanometres. The radius of an atom is around one-tenth of a nanometre (0.1 nm). When particles have nanometre dimensions or even less, they are called *nanoparticles*. Nanosciences and nanotechnologies study, handle and use properties of systems that have dimensions as small as a few atoms, just like nanoparticles.

Applications in this field have now begun to appear, and huge research efforts have been made in the nanotechnologies field. At first sight, nanosciences and nanotechnologies are a natural and necessary extension of works at micrometric scale (that have led to microelectronics and computer sciences), looking to create intensive miniaturization. In fact, it is really an important step up – as important as the step from the Wright brothers' plane to the Apollo spacecraft. To get a clear understanding, let us now look at how the property of matter evolves when we gradually change from our macroscopic scale to the nanometric scale.

1.2   From the macroscopic to the nanoscopic world

When the characteristic dimensions of devices decrease from macroscopic to nanoscopic (a few microns), effects that were preponderant at our scale become negligible, while other effects become very important. For example, the earth's gravity effect is very important at our scale but is negligible at the micrometre scale. It is superficial tensions (atomic interaction force between surfaces) that have dominating effects. We will look at this matter in more depth later. It means that 'classical' arguments based on our experience of the macroscopic world must be changed; it becomes essential to really understand the involved phenomena. Our intuition is often no longer reliable. The limit between macroscopic and microscopic effects is not clear but stands between a few hundreds and a few micrometres.

When we continue decreasing the dimensions to reach the nanometre, another boundary appears. Above micrometres, macroscopic properties of matter are still valid, but this is not true at the nanometre scale. The number of atoms at the surface becomes non-negligible compared with the number in the volume. The matter's behaviour leads to new physical, chemical and biological properties. This *top-down* approach allows us to consider extrapolation methods of microelectronics or chemistry, for instance, at smaller scales. Most of the devices labelled 'nano' on the market nowadays come from this approach.

Going from the macroscopic to the nanoscopic world, we come across an ill-defined area, between the 'classical' and 'quantum' worlds. This field is called *mesoscopic* physics – a term invented in 1976. This area of physics is a few decades old. However, its uses can come to the fore again with the development of micro- and nanotechnologies.

The opposite approach, the *bottom-up* one, lets us understand how to go from the atomic scale to nanosystems. When atoms congregate, they make more or less complex molecules, becoming atom clusters whose shapes evolve into a polyhedron. When the cluster is composed of hundreds of atoms or more, we get nanoparticles that can grow into structures of a few micrometres in dimension.

At the atomic or molecular scale, classical physics is no longer enough to understand the properties of particles: a knowledge of quantum mechanics is required. This is one of the aspects of nanotechnologies that make research investigations simultaneously interesting, disconcerting (particularly for engineers) and difficult.

Theoreticians in chemistry, have for some time known how to determine the properties of molecules, a field known as *quantum chemistry*. However, the field requires the aid of powerful computers to conduct the time-consuming and complex computations. The greater the number of atoms, the more powerful the computer required. Computing the properties of nanoparticles of hundreds of atoms is almost impossible.

At the other end of the dimensions' scale are the solids, which are crystalline and therefore have interesting mathematical properties. Their spatial arrangement is periodic and consequently the calculus can be simplified. This field is called *solid-state physics*.

We can see that nanosystems have properties that are 'halfway' between molecular and solid-state. Unfortunately, simplifications of computations for these two ends are not possible. Therefore a whole new field of research is open for scientists.

1.3   From fundamentals to applications

In the past, fundamental research and applications often happened in different times. In the nanotechnologies domain, this separation period between fundamentals and applications is much reduced. Scientists have been studying entities the size of atoms and molecules for a long time. Atomic and molecular physics, spectroscopy, chemistry, biochemistry and biology are affected by properties of atoms and molecules. However, in these domains such tiny entities could not be handled directly and could not be analysed separately. It is only in the 1980s that nanoparticles were able to be handled separately.

At this scale, it is not possible to distinguish physical and chemical properties of nanosystems, which highly depend on the way they are synthesized, arranged and used. Therefore, physicists, material specialists, engineers and biologists must work together in order to understand and use the properties of nanosystems. Moreover, the development of nanosystems for experimental study requires specific equipments that fall within the competence of applied sciences and engineering. The interaction between sciences and technologies has to be tight. This interaction is between all the players of the different involved fields. However, each field must keep its identity; physicists, chemists, and engineers have their own specific skills and way of thinking. Therefore it would be unrealistic to try to concentrate them in a single specialist; it is more productive to make these different specialists work together on common subjects. This is one of the reasons why people working in theses fields must be multidisciplinary.

What emerges from what has been described earlier is that separation between nanosciences and nanotechnologies does not have a real meaning. That is why most of the time the term *nanotechnologies* also includes nanosciences.

The interest for nanotechnologies comes also from the fact that the potential applications are far beyond the electronics field. In the coming decades, nanotechnologies will radically transform sciences, technology, economy and society. Economists expect that by 2010–15 the annual world turnover of nanotechnologies will be around a thousand billion euros. Therefore, nanotechnologies are considered worldwide to be a strategic domain.

Moreover, in making progress in research, it is noticeable that the fields of MEMS and nanotechnologies are not unconnected. When the size of MEMS is reduced, some elements' sizes are close to what can be found in nanotechnologies; this field is known as NEMS (nano-EMS). Therefore, there appears an increasingly strong complementarity between microsystems and nanotechnologies.

1.4   A different physics

In order to understand how the behaviour of matter is different from the one we know at the macroscopic scale, let us look at how the properties of matter evolve when we go gradually from the macroscopic to the nanometric scale.

From the macroscopic world to nanotechnologies, we cut across two physical limits. The problem is complex because (1) these limits are not clear and (2) they depend on the effects we are looking at, i.e. on the materials being considered. Therefore, prior to any work on nanotechnologies, an understanding on how the properties evolve with the scale of the dimensions is required. For that reason, the resort to the *scale laws* is very useful.

1.4.1   Scale law

We are going to analyse sequentially scale laws related to mechanical engineering, fluids, electromagnetics, thermodynamics and optics. The most important are summarized in Table 1.2.

1.4.2   Mechanics

Let us consider elements whose characteristic linear dimension is L. In the following, except when it is clearly stated, we will consider that all linear dimensions vary proportionally with L. Therefore, all areas vary as L^2:

$$S \sim L^2 \tag{1.1}$$

and volumes vary as L^3

$$V \sim L^3 \tag{1.2}$$

Therefore, masses m vary as

$$m \sim L^3 \tag{1.3}$$

Table 1.2   General scale laws

Quantity or phenomenon	Definition	L^n	Remarks
Gravity force	$F_{gr} = mg$	L^3	—
Ground pressure	$P_{gr} = F_{gr}/S$	L	—
Adhesion force	F_{vdw}	L^2	Van der Waals, Casimir
Macroscopic frictional force	$F_{fr} = \mu F_{gr} = \mu mg$	L^3	—
Microscopic striction force	F_{str}	L^2	—
Kinetic energy	$E_c = mv^2/2$	L^3	v constant
		L^5	$v \sim L$
Potential energy due to gravity	$E_{pot} = mgh$	L^3	h constant
		L^4	$h \sim L$
Moment of inertia	$I = cst \cdot mL^2$	L^5	—
Rotational kinetic energy	$K = (1/2)I\omega^2$	L^5	ω constant
Maximal deflexion of a suspended beam	ζ	L^2	under its own weight
First mode resonant frequency	ν	L^{-1}	Pipe, strips, springs
Speed limit of free falls in fluids	$v_{lim} = 4\rho g r^3/18\eta r$	L^2	—
Damping time	τ	L^2	—
Reynolds number	$Re = \rho v L/\eta$	L^2	$v \sim L$
Diffusion time	$\tau_{diff} = L^2/\alpha D$	L^2	—
Electrical resistance	$R_{el} = \rho_{el}L/A$	L^{-1}	—
Electric current	I_{el}	L	V_{el} constant
Joule effect	$W = R_{el}I_{el}^2$	L	V_{el} constant
Electric field	E_{el}	L^{-1}	V_{el} constant
Capacitor's capacitance	$C = \varepsilon_0 A/d$	L	parallel plates
Capacitor's charge	$Q = C \cdot V_{el}$	L	V_{el} constant
Energy stored	$E_{cap} = Q^2/2C$	L	V_{el} constant
		L^3	Constant charge density
Force between plates	F_{cap}	L^2	—
Magnetic field in solenoids	$B = \mu I_{el}n/L$	L	n constant; constant curent density: $I_{el} \sim L^2$
Magnetic energy in solenoids	$E_{magn} = B^2 \cdot V/2\mu$	L^5	—
Magnetic force	F_{magn}	L^4	—
Thermic energy	E_{th}	L^3	—
Thermal loss: power dissipation	P_{diss}	L^2	conduction, radiation
Thermal constant time	τ_{th}	L^2	—

Let us first analyse how forces vary with L. The most common force is gravitational. At the earth's surface, the gravitational force is $F_{gr} = mg$, where g is the acceleration of gravity:

$$F_{gr} \sim L^3 \tag{1.4}$$

The pressure exerted by a body to the ground is $P_{gr} = F_{gr}/S$, and

$$P_{gr} \sim L^3/L^2 = L \tag{1.5}$$

At the microscopic scale, adhesion forces are dominant (we will go into this in greater depth later). Let us consider first the adhesion of two surfaces that are a distance away from each other. Adhesion between solids is due to the force between atoms and molecules. The main adhesion force is Van der Waals's force when x is between 2 and 10 nm. It is obvious that $F_{vdw}(x)$ is proportional to the contact area:

$$F_{vdw}(x) \sim L^2 \tag{1.6}$$

Since F_{gr} and F_{vdw} vary differently with L, their relative value varies also with L. It becomes

$$F_{vdw}/F_{gr} \sim L^{-1} \tag{1.7}$$

It means that the adhesion force is bigger than the gravitational force (due to the earth's gravitation) when L becomes small. The critical value of L for which the two forces are equal depends on x and on the properties of the medium between the two plates. Below approximately $L = 1$ mm, F_{gr} is smaller than F_{vdw}. From then on, gravitational forces can be neglected at the micrometric scale and below.

When the two surfaces slide against each other, at the macroscopic scale the friction force is given by: $F_{fr} = \mu F_{gr} = \mu mg$, where μ is the friction coefficient. When μ is constant,

$$F_{fr} \sim L^3 \tag{1.8}$$

F_{fr} does not depend on contact surface area. The acknowledged reason is that the two rough bodies are geometrically in contact at only a few points. At the microscopic scale, things are different because of inter-atomic interactions: adhesion forces are huge. Striction forces F_{str} (a combination of adhesion and friction forces) must be considered. They vary proportionally with the contact area; it becomes

$$F_{str} \sim L^2 \tag{1.9}$$

This last relation is appropriate when L is small (a few nanometres) while relation (1.8) is applicable when L is large. At what value of L are the two relations comparable? At present, it is not possible to answer this question because there are too many parameters to take into account, such as the roughness of the surface in contact, or the stress resistance of the material. Nonetheless, solving microsystems problems at the nanometric scale, relation (1.9) is preponderant.

Let us now analyse the case of springs. The restoring force is given by $F_{spring} = -k \cdot \delta L$, where k is the spring constant and δL is the elongation. When k is constant we obtain

$$F_{spring} \sim L \tag{1.10}$$

The corresponding oscillation frequency is: $\nu_{spring} = (1/2\pi)(k/m)^{1/2}$ and

$$\nu_{spring} \sim L^{-3/2} \tag{1.11}$$

Therefore, the oscillation period varies as

$$T_{spring} \sim L^{3/2} \tag{1.12}$$

The kinetic energy of a body, E_c, is given by $E_c = m\nu^2/2$. Then, when ν is constant it becomes

$$E_c \sim L^3 \tag{1.13}$$

When $\nu \sim L$ we obtain

$$E_c \sim L^5 \tag{1.14}$$

The gravitational potential energy, $E_{pot} = mgh$ (when h is constant) vary as

$$E_{pot} \sim L^3 \tag{1.15}$$

When $h \sim L$, it becomes

$$E_{pot} \sim L^4 \tag{1.16}$$

The potential energy of a spring $E_{pot, spring} = -k \cdot \delta L^2$, and

$$E_{pot, spring} \sim L^2 \tag{1.17}$$

Bodies in rotation are characterized by their moment of inertia, $I = constant.mL^2$. Then:

$$I \sim L^5 \tag{1.18}$$

The rotational kinetic energy is $K = (1/2)I\omega^2$, and

$$K \sim L^5 \tag{1.19}$$

This means that, when ω is constant, the rotational energy of small systems decreases rapidly with the size. And in a rotating system the kinetic energy 'stored' is smaller for small systems than for bigger ones.

In nature, we often find strips, springs or pipes characterized by resonance frequencies. The lowest natural frequency, ν, generally matches up with a state where the length of the device is equal to a quarter or half the wavelength, λ. From the relation $\nu = \lambda \cdot \nu$ (where ν is the phase speed of the wave), it becomes

$$\nu \sim L^{-1} \tag{1.20}$$

Resonance frequencies are high in small systems. Often, we can find suspended elements in natural or artificial systems. The deflexion ζ of a suspended strip due to its own weight varies as

$$\zeta \sim L^2 \tag{1.21}$$

Solid materials, when compressed, can withstand a maximum tension before breaking, T_{br}. In the case of animal bones, when they are loaded only by their own weight, mg, then $T_{br} \sim mg/S$, where S is their section. Then

$$T_{br} \sim L^3/L^2 = L \tag{1.22}$$

For a given material, T_{br} is constant. Consequently, when the dimensions change, the minimum diameter d of the supporting material is such that $d^2 \sim L^3$, or

$$d \sim L^{3/2} \tag{1.23}$$

1.4.3   Fluid mechanics

Except under vacuum, all bodies move in a fluid (air, water, etc.), so it is necessary to understand the effect of fluids on the movement. When a body falls vertically in a fluid, because of viscous frictions, after a certain time it will fall at a constant speed, v_{lim}. When the body's shape is spherical, the limit speed is given by $v_{lim} = 4\rho gr^3/18\eta r$, where r is the radius of the sphere, η is the fluid viscosity and ρ is the density (also volumic mass).

$$V_{lim} \sim L^2 \tag{1.24}$$

$$\tau \sim L^2 \tag{1.25}$$

Viscosity forces rapidly absorb all movements when dimensions are small. If the air was still, a very small body would be motionless in the air. At high speed, hydrodynamic effects are not stable and turbulent flows appear. The laminar-turbulent transition is explained by the Reynolds number, $Re = \rho v L \eta$. When $v \sim L$, then

$$Re \sim L^2 \tag{1.26}$$

For flows in pipes, the laminar-turbulent transition occurs when $Re \approx 10^3$. Consequently, turbulent flows disappear in microsystems. A particle travels a distance L by diffusion during a diffusion time given by $\tau_{diff} = L^2/\alpha D$, where α is a geometrical constant and D is the diffusion constant. It becomes

$$\tau_{diff} \sim L^2 \tag{1.27}$$

This relation is valid for particles diffusions and thermal diffusions.

1.4.4   Electromagnetism

A conductor of length L and section A is characterized by an electrical resistance $R_{el} = \rho_{el} L/A$, where ρ_{el} is the electrical resistivity. Then

$$R_{el} \sim L^{-1} \tag{1.28}$$

When a difference of potential is applied, V_{el}, the current I_{el} is given by Ohm's law, then

$$I_{el} \sim L \tag{1.29}$$

Consequently, the power dissipated through the device is given by Joule's law: $W = R_{el}I_{el}^2$, and

$$W \sim L \tag{1.30}$$

The number of elements per unity of surface area varies as L^{-2}. Then, the dissipated power per unity of surface area varies as

$$W_{un} \sim L^{-1} \tag{1.31}$$

It is an acknowledged fact in microelectronics that the reduction of the size of components makes power dissipation increase. A way to reduce this effect is to lower the applied voltage.

When V_{el} is constant while L varies, the electric field E_{el} varies as

$$E_{el} \sim L^{-1} \tag{1.32}$$

When the electric field increases, E_{el} can reach a range where in the semiconductor materials Ohm's law is no longer valid (around 10^7 V/m).

Microsystems can include capacitors. We will see in other chapters that the charge of nanoparticles plays an important part in the transport of electrons through the nanoparticles. Let us consider a capacitor composed of two parallel plates of area A, at distance d from each other. The expression of the capacitance is $C = \varepsilon_0 A/d$, and

$$C \sim L \tag{1.33}$$

When a potential difference V_{el} is applied at the capacitor terminals, a charge Q appears on each plate $Q = C \cdot V_{el}$ and

$$Q \sim L \tag{1.34}$$

Under these conditions, the energy stored in the capacitor is $E_{cap} = Q^2/2C$ and

$$E_{cap} \sim L \tag{1.35}$$

Then the energy stored decreases with the capacitor's size. Sometimes people prefer to work with capacitors that have constant charge density, i.e. $Q \sim L^2$. In this case,

$$E_{cap} \sim L^3 \tag{1.36}$$

The electric force F_{cap} between the plates of the capacitors is given by the derivative of E_{cap} with respect to their distance, therefore,

$$F_{cap} \sim L^2 \tag{1.37}$$

The magnetic field in a solenoid composed of n coils, of length L and with a current I flowing through is given by

$$B = \mu I_{el} n / L$$

When the solenoid gets smaller with n constant, the coils' section is also reduced and I_{el} varies. It is realistic to assume that the current density is constant. Therefore, $I_{el} \sim L^2$ and

$$B \sim L \tag{1.38}$$

The magnetic energy stored in the solenoid is

$$E_{magn} = B^2 \cdot \text{Vol}/2\mu$$

where Vol is the magnetic field volume. Because $\text{Vol} \sim L^3$, it becomes

$$E_{magn} \sim L^5 \tag{1.39}$$

$$F_{magn} \sim L^4 \tag{1.40}$$

When going down to small dimensions, technological issues must be taken into account. For instance, when L is small, the number of coils is not independent from L because we cannot manufacture wires that are too thin. In practice, solenoids smaller than 1 mm^3 are difficult to manufacture. Moreover, the maximum current density is limited by energy dissipation, that is to say the maximum temperature of the system. This dissipation depends on the size of the device. Therefore the current density varies with L, in a way that depends on the fabrication of the device.

1.4.5   Thermodynamics

The energy E_{th} that is needed to heat a system at a temperature T is proportional to the mass:

$$E_{th} \sim L^3 \tag{1.41}$$

Conduction, convection and radiation are the three ways of heat dissipation. Regarding conduction and radiation, the power dissipation, P_{diss}, is proportional to the area of the system:

$$P_{diss} \sim L^2 \tag{1.42}$$

We must be cautious when using this relation, as heat loss is related to the difference of temperature inside and outside the system. The heat diffusion coefficient must be carefully analysed. The time τ_{th} necessary to the temperature homogenization is proportional to the square of the system linear dimensions:

$$\tau_{th} \sim L^2 \tag{1.43}$$

When the size of a system drops off to the nanometric region, the ratio between the number of atoms at the surface and the number inside the system increases. Therefore, the surface affects the properties of the system; we will investigate this matter in more depth in the next chapters.

1.4.6   Optics

When a planar wave is sent to a system of dimension L, the wave reflects and deviates according to an angle $\theta \approx \lambda/L$. Then

$$\theta \sim L^{-1} \tag{1.44}$$

In photolithography, when a surface is illuminated through a lens of a certain numerical aperture ($ON = n\sin\theta$), the minimum diameter of the irradiated region is $L = 2\lambda/\pi ON$. Then, the required wavelength to obtain a pattern of diameter L varies as

$$\lambda \sim L \tag{1.45}$$

That is why in microelectronics, we use the ultraviolet when we need to manufacture very small electronics components.

The angular resolution $\Delta\theta$ of a system (such as the eye) is limited by a circular aperture of diameter L is given by the same relation. Then

$$\Delta\theta \sim L^{-1} \tag{1.46}$$

When lenses have a given shape, their focal length f varies. f is given by the optical formula:

$$(n-1)(R_1 + R_2)f = R_1 \cdot R_2$$

Consequently,

$$f \sim L \tag{1.47}$$

We can deduce then that the minimal diameter of the focal point, $d_f = f\Delta\theta$ doesn't vary with L:

$$d_f \sim L^0 \tag{1.48}$$

This is valid until the theory of Fraunhofer diffraction is applicable, to be precise, while the elements' dimensions are greater than λ.

1.5   Some examples

These scale laws show that physical phenomena vary differently according to the size of elements. In the following we will show their utility via some simple examples.

When we want to compare different physical quantities according to L^n-type laws (for different values of n), we can observe that the bigger n is, the more important is its effect for great values of L. This is illustrated in Figure 1.1.

Let us first consider an example of energy conversion: the steam micro-engine. A small quantity of steam is trapped in a cavity closed by a piston. The system is

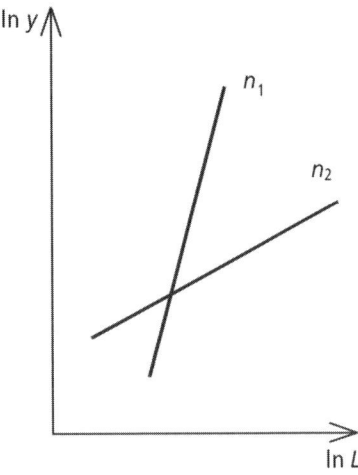

Figure 1.1   Representation of the function $y = a. Ln (n_1 > n_2)$

heated by Joule effect. The generated pressure is enough to drive the piston. Let us compare the potential differentials to apply in order to heat the system; heat losses will not be considered. We assume all the linear dimensions vary proportionally. The thermal energy required to evaporate the water is $E_{th} \sim L^3$, while the energy due to the Joule effect (assuming that the heating lasts the same amount of time) varies as $W \sim L$, when V_{el} is fixed. The amount of W required to evaporate the water is L^2 times the amount we get when we simply reduce all dimensions. Because $I_{el} \sim L$ when V_{el} is constant, the required applied tension varies like L.

Let us consider another example, and assume that the electrical energy stored in a capacitor is fully converted into the rotation of a wheel. How does angular speed vary with L when all dimensions vary proportionally? From (1.19) and knowing that $K = (1/2)I\omega^2$, we can deduce that $\omega \sim L^{-5/2}$. The 'free' rotation is then easier in the micro-world than at our scale. However, friction must be taken into account as well. With small dimensions, $F_{str} \sim L^2$ (1.9), while $F_{fr} \sim L^3$ (1.8) at the macroscopic scale. Therefore, the dissipated power due to friction P_{fr} is proportionally more important in the micro-world. Assuming that the linear speed is proportional to L, then $P_{fr} = F_{fr}$. $v \sim L^4$ is proportional for large dimensions and $\sim L^3$ for small dimensions. A lot of examples can be analysed in the same way.

From all of this, we can conclude that interpretations and intuitions based on our everyday experiences at our own scale must be reconsidered when dealing with the micro-world. Things get worse when we go down to the nano-world, where quantum effects are dominant, as we shall see in the following chapters. Extrapolations to small dimensions, of models used at our scale, are usually not applicable. Because of the natural complexity of the fabrication of micro- and nanosystems, it becomes obvious that numerical simulations can and will play an important part in the study and the development of these technologies.

Table 1.3   Forecast of the world annual nanotechnolo-
gies market for 2010–15 (in €bn)[Source:
www.nano.gov]

High-performance and nanostructured materials	340
Electronics	300
Pharmacy	180
Chemical industry	100
Aerospace industry	70
Power saving in lighting	100

1.6   Various applications

As we mentioned earlier, research into these technologies is worth pursuing because the application fields of nanotechnologies are so numerous. Researchers will be able to develop structures that do not exist in nature, beyond what chemistry can offer.

Among the advantages of these materials and devices, we can notice light materials, that are more resistant, programmable; cuts in life-cycle costs through a reduction in breakdowns; new appliances based on novel principles and architectures; the use of manufacturing processes at the molecular and cluster scale; and so on.

From nanoelectronics to biotechnology, from space applications to the protection of the environment to petrochemistry or to the development of materials with specific properties, the possibilities are enormous. A forecast of the nanotechnologies market for 2010–15 is given in Table 1.3.

1.6.1   Nanoelectronics

As we mentioned earlier, miniaturization of microelectronics is limited. If we want the trend to continue, we must find something else. This is the field of nanoelectronics coupled with molecular and quantum computers. Behind this single word lie many concepts with the common prefix 'nano'.

It has been about 30 years since chemists and physicists designed, sketched and produced molecules that individually fulfil functions of resistors, diodes, transistors and switches. In the early 1970s, researchers at IBM patented the idea that a single molecule could work as a memory element. A few years later, researchers imagined that a molecule could behave like a diode when placed between two metallic electrodes. Moving theory into practice was not easy but, thanks to the invention in 1981 of the Scanning Tunnel Microscope (STM), things began to advance rapidly. Attempts of interconnections on a molecule, possibilities of manipulating atoms, measuring the electrical resistance of a molecule – all of these were essential experimental steps, and all achieved thanks to the original work of STM. Nevertheless, various difficulties appeared that, today, make the approach far from the technological achievement we expected in 1980.

In a similar vein to this approach, works on the electronic structure of the materials of polymers show that they can be used in quantum computers. Some also imagine completely different systems based on the manipulation of DNA fragments. The field of molecular and biomolecular computers is currently booming.

Simultaneously, some imagine using particles bigger than molecules of nanometric size (around a few nanometres): quantum dots. In this field, nanoparticles have specific electronic properties, as we will see in Chapter 3. Quantum dots could be used as single electron memory elements. Nonetheless, besides physicochemical problems related to their production and their arrangement, fundamental science issues must still be solved. Other prosaic issues will be raised, for instance, due to the fact that electronic devices dissipate heat. The smaller the device, coupled with the increase in the numbers of the devices, the more important it becomes to tackle heat dissipation, which interferes with the running of a computer.

Even if it seems that the existence of this kind of computer is still some way off, it is still worth working on it. Indeed, miniaturization is generally associated with a high computing power in a small volume, which is essential if we want Moore's law to be verified beyond 2010.

1.6.2   Biotechnologies

The fundamental molecular components of life (proteins, nucleic acids, lipids, etc., and their non-biological kind) are materials that have exclusive properties, fixed by their size, shape and arrangement at the nanometric scale. Biosynthesis and bioprocess engineering provide new methods of manufacturing chemical and pharmaceutical products. The integration of fundamental biological components in materials and synthetic devices will allow the combination of biological functions with other selected properties of some materials, as we already see with composite materials that mix mechanical properties with electrical ones etc. The imitation of biological systems is an important research activity in many fields. For instance, the research activity in biomimetic chemistry is based on such an approach. It is also worth mentioning again the laying-down of some selected biological structures over pre-designed substrates, or the processing of molecules at nanometric scale.

1.6.3   Biomedical field

Living systems are ruled by the behaviour of molecules at the nanometric scale. Recent works in the nanofabrication field suggest that the actual complex process of genome sequencing and detection of gene expression can be made easier and faster using nanoscopic surfaces and systems. On top of making the use of optimal medicines easier, nanotechnologies would offer new injection methods for medicines, thereby extending the potential of therapeutic medicines.

The increase in the possibilities of nanotechnologies should also influence fundamental research on biological cells and pathologies. Due to the development of techniques able to analyse the world at the nanometric scale, we will be able to characterize physical and chemical properties of cells (including cell division and cell locomotion) and measure the individual properties of molecules.

The synthesis of high-performance biocompatible materials should also be possible because of the ability to act at nanostructure scales. Inorganic or organic nanometric materials could be injected into specific places for diagnosis or as active elements. As for the increase in the computing power in 'nanocomputers', it should allow for improved simulations of living systems as well as the development of new biocompatible implants and the monitoring of the auto-dispensing of medicines in specific places.

Nanotechnologies also include nanomagnets that are used particularly in magnetic resonance imaging (MRI). Progress on the reproducible synthesis of nanomagnets should create an increase in resolution that will lead to better diagnoses and healing.

Even if the issue remains open, we hope that the impact of new biomedical techniques on healthcare and their cost advantage for the community will be positive.

1.6.4 Space domain

The space domain is concerned with many aspects of nanotechnologies. One of the essential characteristics of space technologies is the rush to make equipment smaller and lighter. The very high cost of launching a space mission means that the ability to add more while decreasing the weight is becoming essential. Spacecraft must be autonomous over long periods of time. Manned space missions require important safety devices, so it is not surprising that this area is highly concerned with nanotechnologies.

The main fields of interest are nanoelectronics, sensors, nanomachines and materials. In future missions, spacecraft will be autonomous and smart, requiring computers that are powerful and compact, that operate using very low power (particularly for missions to faraway planets), and that are resistant to cosmic radiation. Furthermore, the computing power of onboard computers must be very high in order to assess flight data, apprehend structural aspects, understand the environment and react appropriately. Molecular and quantum computers that resort to nanotechnologies potentially fulfil these requirements.

In space, sensors of all kinds are required. We need to be able to sense immediately and analyse all kinds of things, including temperature, pressure, particles, chemical materials and deformations. The further away the mission, the more resistant the sensors must be to harsh environments, and they must also be light and compact. Integrated nanosensors need to be developed; another advantage is their low power consumption.

Miniaturized elements, smaller than micromachines, could also come into existence. We think about nanoengines, nanopumps, nanothrusters based on micromachine devices that would have been intensively miniaturized. Experiments showing rotating nanosystems, for instance, have already been successfully conducted.

Some materials, including nanoparticles such as composite or multifunctional materials, are also under investigation for the next generation of spacecraft.

1.6.5   Sustainable development

One of the main characteristics and driving forces behind nanotechnologies is the research in industrial processes and the development of environment-friendly technologies. In the field of industrial processes, the fabrication of nanosystems should be realized at the atomic scale or at least at the nanometric scale. Consequently, on the one hand, the quantity of materials required for the production of different nanotechnologies would be automatically minimal. Organic solvents and other hazardous chemical products should be used less and less often – processes that do not rely on such solvents will eventually be developed for specific applications. On the other hand, we hope that the energy required for their fabrication will be less than the energy required today. With elements being smaller, the energy for their manipulation would therefore be reduced.

In the chemical industry, the development of molecular or nanoscopic materials (such as catalysts) should prove to be advantageous, especially by reducing the required material quantity. For instance, in a heterogeneous catalytic reaction, reactions generally take place at the surface of a solid catalyst. Therefore, we would be well advised to increase the surface for a constant volume, i.e. a constant mass. By using nanoparticles, we precisely increase the useful surface.

But it is particularly at the point of use level that energy saving would be noticeable. When carrying out a certain function, the energy required to make a device work decreases rapidly with the size. For instance, spinning a disc at a certain angular speed requires a hundred thousand times less energy when its dimensions are reduced by a factor of ten. Therefore, we would be advised to miniaturize technologies as much as possible; however, to do this, an energy source at this scale must be developed.

For safety, polymers reinforced with nanoparticles can replace metallic structures in car parts, reducing weight and energy consumption. It has been demonstrated that polymers that include nanoparticles are more heatproof, therefore offering benefits and increasing safety in many areas, e.g. the construction industry.

The development of various sensors or machines at nanometric scale cannot be achieved without progress in artificial-intelligence machines. There is no point in developing miniaturized sensors if the command and data-processing systems are not miniaturized. Consequently, we could put various sensors in places we cannot imagine today, to detect such things as deformations, vibration or electric fields. The knowledge of such parameters should allow having information that will let us react more quickly, for instance, in the case of the corrosion of car bodywork. Examples of potential advantages of nanotechnologies are numerous.

In the sustainable-development field, nanotechnologies generate turnovers (in research, development and marketing), and also allow savings to be made (in energy, pollution, healthcare, etc.).

Nanotechnologies are not limited to nanoelectronics: many branches of industry are interested in them. This is probably their essential contribution, as underlined by Tim Harper, founder of CMP Cientifica: 'Their development is the result of the union of various scientific fields and the ability of scientists, in front of complex issues, to share their skills.'

Chapter 2
Atomic structure and cohesion

Particles of between 1 and 100 nm have specific physicochemical properties. These *nanoparticles* are in an intermediate state between solid and molecule. In nanoparticles, the number of atoms at the surface, N_s, is not negligible compared with the number of atoms in the centre of the particle, N_{part}. Consequently, a spherical particle with a radius equal to r interatomic distances, is characterized by the ratio $N_s/N_{part} = 3/r$. A particle of radius $r = 50$ (i.e. around 10 nm), therefore, contains 0.5 million atoms, of which 6 per cent are located at the surface. In those conditions, it is obvious that surface plays an important part in many of the properties of nanoparticles. Given that experimental and theoretical methods for the control of the size and the shape of nanoparticles exist, or are under development, it is easy to guess that we are heading towards the synthesis of particles with selected specific properties.

The specific properties of nanoparticles are a consequence of miniaturization. Since Haüy, at beginning of the nineteenth century, we know that we can describe crystals as a periodic assembly of microscopic entities (that Haüy called 'integrant molecules'). The geometrical shape of these 'integrant molecules' and the way we part the assemblies using plans (crystallographic faces) determine the crystal shape. Today, we know that these 'integrant molecules' are themselves an assembly of atoms. Consequently, there is a correlation between the shapes of crystals and their atomic structure. Each crystallographic face is characterized by an arrangement of atoms, which determines various properties such as the superficial tension or the chemical reactivity of the crystal. When the size drops down to the nanometre scale, specific geometrical structures appear: nanotubes, compact clusters or vacuums (fullerenes), regular polyhedrons.

At this scale, quantum effects are dominant and depend on the arrangement of atoms (as in molecules), but they depend also on the size and shape of nanoparticles. In 1967, Ino and Owaga demonstrated that nanoparticle shapes and structures can be different from crystal ones. The principal reason is the part played by surface. The interaction forced between atoms at the surface is different from those inside, which consequently affects the cohesion of the particles.

In the crystalline state, atom positions are not strictly fixed. The thermal turbulence leads to a synchronized vibration around their equilibrium position. When we reach the fusion temperature, movements become uncoordinated and a phase shift appears. From Pawlov's research in 1909, we know that fusion temperature decreases with the nanoparticle size. This variation is a consequence of the influence of the surface over the cohesion of the nanoparticle. Since then, research has shown that the phase diagrams of nanoparticles are different from solid ones. In certain cases, crystallographic structures are not present in nanoparticles, while in other cases, meta-stable structures in solid state are stable in nanoparticles.

When nanoparticles' size decreases down to the nanometre scale, their shape can vary. This is due to the fact that energy can have many local minima corresponding to different atomic arrangements. A small excitation – caused by an electronic microscope beam, for instance – can be enough to provoke transitions in the nanoparticle structure.

In many applications, it is necessary to stabilize nanoparticle atomic structure. For instance, electronic and optical properties of a nanoparticle's semiconductor depend on the crystalline structure and shape. For nanoelectronic applications, it is essential to work with particles that have a stable and well-defined shape and size. In some cases, the structure can be stabilized by passivating the surface with an – organic or inorganic – external layer.

The issue of the structural stability of nanoparticles is essential for the future development of nanotechnologies. In this chapter, we will first remind ourselves of the principal characteristics of surfaces of solids. This will help us understand their role in the cohesion of nanoparticles.

The transition from solid state to nanoparticle is not abrupt. To verify it, we will first use the thermodynamic approach, in which we suppose that the macroscopic concept of superficial tension remains valid. This *top-down* approach is interesting because it allows us to understand the essential factors of the cohesion in nanoparticles, free or coated, or even constrained when they are encapsulated in a solid matrix. Thermodynamics is valid when the number of atoms in the nanoparticle is high, but it is not valid any more for small particles.

With nanoparticles being halfway between solids and molecules, we will follow the opposite methodology, going from the atom to the nanoparticle: the *bottom-up* approach. This will introduce the problems of clusters, fullerenes and nanotubes. Shapes of atom clusters can depend on the number of atoms they contain.

2.1   Surfaces and interfaces

In nanoparticles, the number of atoms at the surface is comparable to the number of atoms in the heart of the nanoparticle. Consequently, it is essential to understand the concepts related to the surface and the interfaces, before analysing nanosystems in detail.

Although the science of surfaces is not new, it is still important because every solid or liquid is in contact with the outside via its surface. The existence of surfaces locally modifies the property of the matter and it is through the surface that matter interacts with its environment, hence the interest in surface and interface sciences in many domains, such as electronics, heterogeneous catalysis, detergents, corrosion, photography, lubrication and biology. With nanotechnologies, the science of surfaces gains a new interest.

It was previously mentioned that the number of atoms at the surface of nanoparticles is not negligible compared with the entire number of atoms. The specific surface area (total surface area per unit of mass) in the nanoparticle is then very high. Then, nanoparticles usually develop from a few atoms that are themselves often deposited over a surface. The method of developing nanoparticles is based on surface phenomena.

Rarely isolated, nanoparticles are more or less closely in contact with their environment. It is once again a phenomenon related to surfaces and interfaces. Finally, the physical and chemical properties of the nanoparticles are responsible for the various interactions at the surface, hence the importance of a good knowledge of surface and interface properties.

2.1.1   Superficial tension

One of the fundamental concepts of the science of surfaces is the concept of superficial tension or surface free energy, γ. The interatomic bonds are affected by the creation of a surface; some of them are suppressed and others are modified. The result is an increase in the average energy of atoms that are close to the surface. The superficial tension is defined as the energy required to create a surface of area A, at temperature T, volume V, and constant chemical potential μ:

$$\gamma = (\partial H / \partial A)_{T,V,\mu} \tag{2.1}$$

where H is the Helmholtz free energy of the system under consideration.

We can also describe the superficial tension as being a force per unit of length. These two complementary definitions can be used independently.

To the first approximation, the superficial tension is proportional to the number of interatomic bonds broken at the surface under consideration, multiplied by the number of atoms per unit of area. In the case of a crystalline solid, the number of broken bonds depends on the crystallographic face under consideration.

To create a surface, work needs to be performed. The work required to separate two parts of a same material and obtain a surface area that is twice the unit area is called the work of cohesion, W_c:

$$W_c = 2\gamma \tag{2.2}$$

The work required to separate two materials A and B and obtain two surfaces of unit area A and B is the work of adhesion, W_{ad}:

$$W_{ad}(AB) = \gamma_A + \gamma_B - \gamma_{AB} \tag{2.3}$$

where γ_A, γ_B, γ_{AB} are respectively the superficial tensions of A and B, and the interfacial tension between A and B. Equation (2.3) is called the Dupré equation.

2.1.2   Crystals' shape

The superficial tension varies with the orientation of the crystallographic face of crystals. This anisotropy is such that the shape of the crystal at the equilibrium is

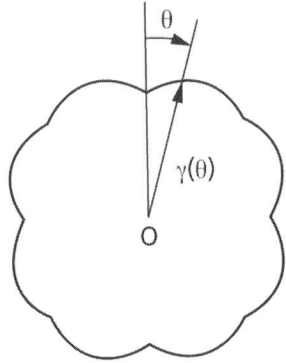

Figure 2.1   Example of γ-plot

controlled by the various superficial tensions of the different crystallographic faces. The shape of crystals is important in nanotechnologies. The anisotropy of crystals can be analysed via what we call a 'γ-plot' obtained as described below.

From an arbitrary origin, O, we draw a vector in the direction n (defined by its polar angle θ, and azimuthal angle Φ), of a length equivalent to the energy of the free surface, $\gamma(n)$, for a surface perpendicular to n. The non-sphericity of the γ-plots shows the anisotropy of γ. $\gamma(n)$ have a minimum in direction n_0 corresponding to the most dense surfaces. Surfaces corresponding to directions that are close to n_0 show a sequence of steps and plateaux. If we call β the energy-per-length unit of a step, we obtain in the region of n_0:

$$\gamma(n) = \gamma(n_0) + \beta|\theta|/d \tag{2.4}$$

where θ is the angle between n and n_0, and d is the distance between the planes along n_0. $|\theta|/d$ is the density of the steps. Consequently, $d\gamma/d\theta$ have a discontinuity at $\theta = 0$, and the γ-plot shows cusps (singular points on the curve) in directions corresponding to n_0, as illustrated in Figure 2.1.

For a crystal limited by a surface S, the shape at the equilibrium must minimize the surface energy, i.e.

$$F_s = \iint \gamma(n)dS; \text{ at constant volume} \tag{2.5}$$

When γ is isotropic, the equilibrium shape is obviously a sphere. When γ is anisotropic, the equilibrium shape is no longer a sphere, but a polyhedral geometrical shape. The exact shape is obtained by the method known by crystallographers as *Wulff's construction*. A two-dimensional example is shown in Figure 2.2.

When the temperature increases, thermal fluctuations appear. The γ-plots show fewer sharp angles. Wulff's construction becomes less pointed and the surface can become rough.

Wulff's construction can be applied to microscopic crystals; it can be applied as well at the scale of nanoparticles.

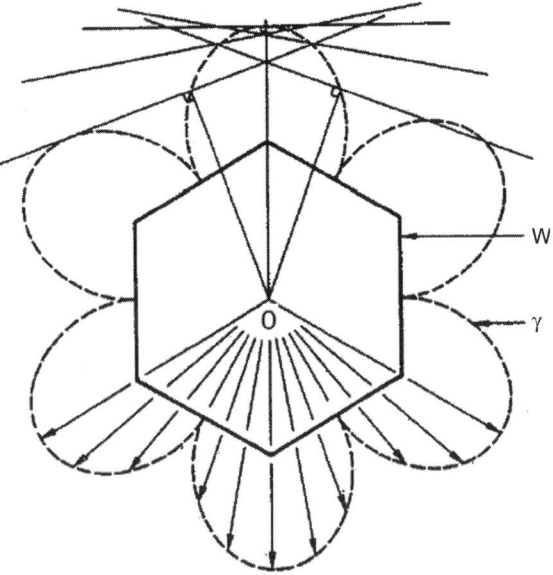

Figure 2.2   Wulff's construction showing how to determine, in two dimensions, the crystal shape from the values of γ

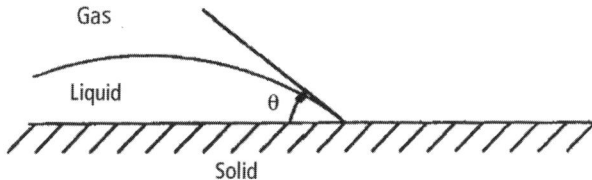

Figure 2.3   Contact angle of a drop

2.1.3   Drops and contact angles

Often, crystals are not isolated, but laid on top of a substrate, on a surface. In this case, the angle between the substrate and the outside face of what is deposited depends on the values of the different superficial tensions.

Let us consider θ, the contact angle (as illustrated in Figure 2.3). The relation between θ and the superficial tensions of the materials in contact, is given by the Young equation:

$$\gamma_A = \gamma_{AB} + \gamma_{AB} \cos \theta \tag{2.6}$$

When $\theta \neq 0$, a drop is formed on the substrate. When $\theta = 0$, it spreads and a monomolecular layer is formed.

To describe the spreading phenomenon, the spreading coefficient is usually defined, σ:

$$\sigma_{AB} = \gamma_B - \gamma_A - \gamma_{AB} = \gamma_A(1 + \cos\theta) \qquad (2.7a)$$

$$\sigma_{AB} = W_{ad}(AB) - W_c(A) \qquad (2.7b)$$

When $\sigma_{AB} > 0$, there is spreading of A on substrate B. When $\sigma_{AB} < 0$, a drop is formed on the substrate.

2.1.4   Development of films on top of a substrate

In nanotechnologies, many techniques used for the synthesis of nanosystems involve a development over a substrate. The issue of film development dates back to 1920, but interest in it has never waned. The advent of nanotechnologies depends on these works as well.

When atoms interact with a substrate, a film is formed over the substrate. There are three different types of growing mechanism, named after their principal investigators: (1) Volmer–Weber mechanism, also known as mechanism of 'island growth'; (2) Frank–Van der Merwe mechanism, also known as 'layer-by-layer' growth; (3) Stranski–Krastanov. These mechanisms are illustrated in Figure 2.4.

The Volmer–Weber mechanism is observed when the smallest nuclei are formed (also called *nucleation*), and grow to become islands; this happens when the deposited atoms or molecules are more bonded with one another other than with the substrate. It is the case with a lot of metals on isolated substrates, or on graphite or mica.

The opposite mechanism (Frank–Van der Merwe) occurs when deposited atoms or molecules interact more with the substrate than between them. Consequently,

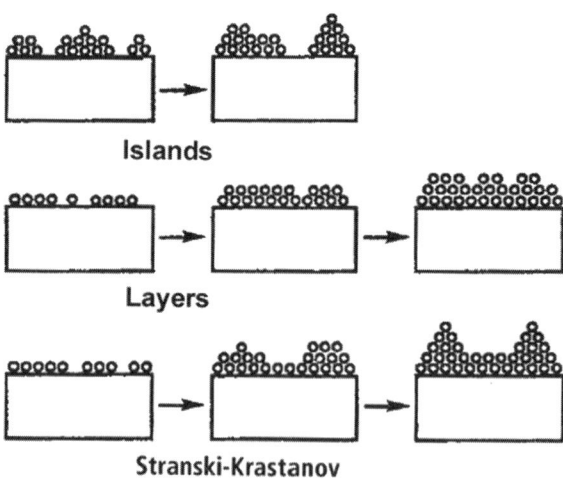

Figure 2.4   Growing mechanisms of thin films over a substrate

successive layers are formed. This mechanism is particularly important in the epitaxial growing of semiconductor materials.

The Stranski–Krastanov mechanism is midway between the two previous mechanisms. After the formation of one or a few monomolecular layers, two-dimensional growth is replaced by three-dimensional growth with development of islands. This can happen when some stresses appear during the epitaxial growing of the first layers on the substrate, particularly due to discordances between the crystalline cells. The stresses first accumulate, and then are cleared with the three-dimensional growth of islands. These common growing methods are found in metal–semiconductor and metal–metal systems.

The growing mechanisms of films are very important for the growing of nanoparticles and other nanosystems. They also occur when atoms gather along a crystallographic step, during the synthesis of nanowires, for instance.

2.1.5   Adhesion phenomena

At the atomic scale, in the previous section, the direct interaction between atoms in contact plays the preponderant part. When atoms or molecules are separated by several interatomic distances, other phenomena occur; these phenomena are useful to describe the adhesion in nanoscopic systems. At this scale, the adhesion forces are greater than the earth's gravity. At the sub-micron scale, two principal kinds of interaction are important, on top of the usual interatomic interactions: the Van der Waals interaction and the Casimir effect.

2.1.5.1   Van der Waals interaction

The adhesion between a solid and another solid (or a liquid) is due to interatomic forces. The principal forces that are responsible of the adhesion are the Van der Waals forces. The Van der Waals potential of interatomic interaction is given by

$$\varepsilon = -C/r^6 \tag{2.8}$$

where C is a positive constant and r is the interatomic distance.

The interaction between two plates, separated by a distance x, is obtained by a sum over all the atom–atom interactions. We obtain

$$\varepsilon_{\text{plate-plate}} = -\pi n^2 C/12x^2 = -H/12\pi x^2 \tag{2.9}$$

where n is the atomic concentration and H is the Hamaker constant. For solids, H is around 10^{-19} J in the air, and 10^{-20} J in water.

The attraction force between the plates is given by the derivative of $\varepsilon_{\text{plate-plate}}$ with respect to x:

$$F(x) = H/6\pi x^3 \tag{2.10}$$

The exact value of $F(x)$ depends on the shape of the two entities interacting. A few examples are summarized in Table 2.1. The Van der Waals interactions occur when the distance between the two entities is less than about 10 nm.

Table 2.1   Van der Waals interactions for some entities

Entities	Van der Waals interaction
Two infinite plates	$-H/12\pi x^2$
Two plates of thickness d	$-(H/12\pi)\,[x^{-2} + (x + 2d)^{-2} - 2(x + d)^{-2}]$
Two spheres of radius r, where the distance centre to centre is equal to $R = s \cdot r$	$-(H/6)\,[2(s^2-4)^{-1} + 2s^{-2} + \ln(s^2 - 4) - \ln(s^2)]$

2.1.5.2   Casimir effect

Another effect that can be important is the Casimir effect, which results from the quantization of the electromagnetic field. Quantum electrodynamics establishes that, for each mode of the radiation field, associated energy levels are spaced out, such as those for harmonic oscillators. Each level is related to a whole number of photons. The quantum analysis of the harmonic oscillator shows that the fundamental level has a finite energy, not equal to zero. Regarding electromagnetic fields in 'free' space, there is an infinite number of radiation modes. In the presence of two parallel plates, completely conductive, only the longitudinal waves with nodes on each surface are present. The net result is that the fundamental energy is modified. This energy modification is shown as a fall of the potential energy when the plates become closer, resulting in an attraction force between the conductive plates.

This force is proportional to the plates' surface areas. The force per unit area (i.e. pressure) between two 'infinite' conductive plates separated by a distance x is given by

$$P = \pi\,\hbar c/240x^4 \tag{2.11}$$

For $x = 10$ nm, $P = 1.25$ atm. It is then an important effect. When the plates are not completely conductive, P must be modified by a factor that depends on the dielectric constant of the materials.

2.1.6   Adhesion work

We have previously defined the adhesion work between two surfaces, according to the superficial tensions of the related surfaces (Dupré equation). The adhesion work is also the work done by the attraction forces between surfaces.

The adhesion force is useful because it distinguishes two states: contact and separation. This work is done over a short distance. Regarding Van der Waals forces, 99 per cent of the work is done when the surfaces are at 1 nm from each other. For other type of bonds (ionic, covalent), the distance is even smaller. Consequently, the precise shape of $F(x)$ is not essential to understand a lot of phenomena, because of the instability due to the attraction that often cannot be measured.

The adhesion work allows us, thanks to various methods, to calculate the mechanical energy required to separate two surfaces. Consequently, to pull away two identical spheres of diameter D and of adhesion work W, a force F is required:

$$F = kWD \tag{2.12}$$

where k is a constant close to one. Regarding elastic spheres, $k = 3\pi/8$. This relation is a useful link between mechanics (via F) and chemistry (via W). By applying the same methodology to different geometries, we can deduce the adhesion equations that describe the mechanism of adhesive joints.

This link between adhesion work and separation force of surfaces allow understanding of different effects, such as the gluing. For instance, a layer of liquid spread on a surface modifies its superficial tension γ; therefore, it modifies W also and F. If the liquid is wetting, γ decreases, therefore W also decreases as well as F. If the liquid is not wetting, γ increases.

Those effects play a part as well, when we deal with systems at the micrometre scale or below, because in this region the adhesion is dominant compared with the gravity. For instance, regarding nanomachines, it can be difficult to pull away two moving surfaces when they have come into contact.

2.2 Thermodynamics of nanoparticles

When we want to understand the origin of the properties of nanosystems, the easiest approach involves analysing how their size and shape affect their properties. Going from the solid state to the nanoparticle (the top-down approach) is, historically, the first approach followed. The advantage is that we can start from established concepts: those of thermodynamics. The thermodynamics analysis, in addition to its simplicity, allows for the demonstration of other important effects, such as the variations of the phase diagram with the size of nanoparticles, their shape, defaults they have, or stresses they are subjected to.

2.2.1 Thermodynamics description

One of the conditions of applicability of thermodynamics is that the number of atoms involved is 'high enough'. What does 'high enough' mean? Let us consider a cube of side length L with N atoms per unit volume. The relative temperature fluctuation $\delta T/T \approx (NL^3)^{-1/2}$, then $L \approx (\delta T/T)^{-2/3} N^{-1/3}$. Let us assume that the temperature is uniform when $\delta T/T$ is inferior to 10^{-3}. In solids and liquids, $N \approx 10^{23}$ cm^{-3}, and $L \approx 20$ nm.

2.2.2 Temperature definitions

In thermodynamics, temperature is the basic parameter. Therefore, it is useful to ask ourselves the fundamental question: what is temperature? For common systems, it is easy to define a local temperature. However, what is the size of the region in which we can define a local temperature? Three different ways of defining a local temperature exist.

The first one is used, for instance, in molecular dynamics. In this method, we calculate the position and the speed of each atom at every time step. Then we calculate the average kinetic energy, E_c. The average must be calculated over a long time period in order to obtain a result that is statistically valid. The calculi are purely classical. The temperature is given by the classical expression:

$$\langle E_c \rangle = \langle mv_i^2 \rangle / 2 = 3kT/2 \tag{2.13}$$

The temperature is defined locally, on an atom. This approach neglects quantum effects.

The second definition of temperature takes into account quantum effects. The collective movements of the atoms are described from the phonons theory. Phonons are characterized by their pulsation $\omega(p, q)$, which depends on the wave vector q, and the polarization p. In quantum mechanics, the average kinetic energy is given by

$$\langle mv_i^2 \rangle / 2 = \sum \{\hbar\omega(q) / [\exp(\hbar\omega(q)/kT) - 1]\} \tag{2.14}$$

At high temperature (far above Debye's temperature), classical and quantum equations give the same results. Below or around Debye's temperature, the two definitions give different temperatures.

A third definition consists of subtracting the movement at 0 K from the quantum definition. The discussion of the previous case remains valid.

Which of these three definitions is correct? It depends on the size of the region where T is defined. The classical definition is purely local. T can be defined for each atom or for each row of atoms.

In quantum definitions, the scale of length is defined from the mean free path of phonons, l_{ph}. At an ambient temperature, the phonons' mean free path in insulating materials is around a few nm (2.3 nm for NaCl, 4 nm for quartz). If two regions are characterized by two different temperatures, the distribution of the phonons will be different. A region is defined by a phonons' distribution. Consequently, the characteristic dimension of a region at a constant temperature must be greater than l_{ph}. But l_{ph} depends on the phonons' frequency. Phonons with low frequencies have a mean free path longer than phonons with high frequencies. At high temperatures, we can define a mean free path. In these cases, the temperature cannot be fixed on a single atom or a row of atoms.

For most materials at an ambient temperature, l_{ph} is in the region of the nm. Consequently, T is defined when the particle dimensions are in the region of the nm.

2.2.3   *Nanoparticles' energy*

The thermodynamic description of nanoparticles is based on the estimation of their free energy, $G(T)$. Let N be the number of atoms in the particle. It is necessary that N be high enough in order for the thermodynamics theory to be valid. We also assume that the particle's surface is characterized by only one value of superficial tension. At a given temperature, Gibbs free energy of a free particle is given by

$$N.G = N.G_\infty + fN^{2/3}\gamma \tag{2.15}$$

where f is a geometrical term that depends on the particle's shape. The term $(fN^{2/3})$ is equal to the number of atoms at the surface. Γ is the superficial tension per atom, that is to say the superficial tension divided by the number of atoms at the surface. For most inorganic materials, γ is practically independent of the temperature T. G and G_∞ are, respectively, the volume energy per atom, considering the particle or the volume material, in a given phase.

The stability of a certain phase compared with another one is the result of the minimization of the free energy between the considered phases. Let G_i, $G_{i\infty}$ be the free energies per atom of the particle and of the volume; and γ_i the superficial tension in the phase $-i$. The equilibrium between the phases satisfies:

$$N(G_1 - G_2) = N(G_{1\infty} - G_{2\infty}) + fN^{2/3}(\gamma_1 - \gamma_2) \tag{2.16}$$

The phase transition happens when $(G_1 - G_2) = 0$. Phase 1 is the phase that is thermodynamically stable when $(G_1 - G_2) < 0$. When we take into account the surface term, it is obvious that the temperature transition of the nanoparticle does not correspond to the volume one. The temperature depends on the particle's size.

2.2.4   Fusion of spherical nanoparticles

Among the various phase transitions, fusion has been known for a long time. In the case of nonorganic solids, the fusion temperature T_m is 'far superior' to Debye's temperature of the solid. Consequently, the calorific capacity is almost constant, and

$$G_{1\infty} - G_{c\infty} = C - B \cdot T \tag{2.17}$$

where C and B are constant for a given material. The suffixes 1 and c refer respectively to the liquid and the solid.

In most inorganic solid materials, γ is almost independent of the temperature. In these conditions, the nanoparticle's fusion temperature varies with its radius R, as

$$T_m = T_{m\infty} + f(\gamma_1 - \gamma_c)/B \cdot N^{1/3} = T_{m\infty}[1 - \alpha/(2R)] \tag{2.18}$$

where $T_{m\infty}$ is the common fusion temperature. The linear variation of T_m with R^{-1} has been shown experimentally in the case of metals and semiconductors. The nonorganic materials are characterized by a constant α positive, between 0.4 and 3.3 nm. Some characteristic values are summarized in Table 2.2.

Many expressions of the parameter α, based on various models, have been proposed. The best known is the original expression of Pawlov (1909), reassessed by Hanszen in 1960:

$$\alpha = 4V_s[\gamma_{sv} - \gamma_{1v}(\rho_s/\rho_l)^{2/3}]/(H_mR) \tag{2.19}$$

where V_s is the molar volume of the crystal, γ and ρ are respectively the interfacial energy per unit area and the volumic mass; the suffixes s, l and v describe the different solid, liquid and vapour phases, H_m is the molar enthalpy of fusion.

Table 2.2   *Common fusion temperature ($T_{m\infty}$) and coefficient of variation of the fusion temperature of different materials.*

Material	$T_{m\infty}$ (K)	α (nm)	α (nm)
Ag	1234	1.27	
Al	933	1.14	0.6
Au	1336	0.92	0.96
Co	1768	1.00	
Cr	2148	1.05	
Cu	1356	1.02	
Ge	1210.6	2.30–3.33	
In	429.4	1.95	0.974
Mo	2883	0.98–1.58	
Pb	600.6	0.98–1.40	1.048
Pd	1825	0.88–1.43	
Si	1683	1.88	
Sn	505.1	1.57	1.476

For most cubic metals, $\gamma_{sv} - \gamma_{lv} \approx \gamma_{sl}$, and $\rho_s \approx \rho_l$. Consequently,

$$\alpha = 4 V_s \gamma_{sv}/(H_m R) \tag{2.20}$$

By introducing (2.20) in (2.18) we obtain the Gibbs–Thomson equation.

In previous equations, we examined the role of interfacial tension. As soon as the particle is surrounded by a layer different from the one at the heart (for instance, an oxide at the surface of a metallic particle), or when the particle is confined in a matrix, the interfacial tension is different from that of the free particle. This has been demonstrated in the case of lead or indium's particles in an atmosphere that contains oxygen. In these cases, the variation of T_m with R is less important compared with the variation in the case of a pure particle.

When the particle is confined in a matrix, the variation of T_m is given by

$$T_m/T_{m\infty} = 1 - [3V(\gamma_{sm} - \gamma_{lm})/R - \Delta E]/H_m \tag{2.21}$$

with $V = (V_s + V_1)/2$, where V_1 is the molar volume of the liquid. γ_{sm} and γ_{lm} are the interfacial energies between the solid and the matrix and between the liquid and the matrix. ΔE is the energy density difference between the solid and the liquid. When ΔE is negligible, T_m can be inferior or superior to $T_{m\infty}$ depending on the sign of $(\gamma_{sm} - \gamma_{lm})$.

2.2.5   Fusion of non-spherical nanoparticles

Nanoparticles do not always have a spherical shape. For instance, when they are synthesized by laser pulse irradiation of a metallic target, metallic nanoparticles in cylinder or plate form can be produced. In this case, at constant volume, the area of the surface is larger compared with spherical particles. Consequently, the variation of T_m with the size of the particle is of more importance when compared with the spherical case.

In the previous equations, we have assumed that each different phase is characterized by a single superficial value. When the size of the nanoparticles decreases to under 2 nm, they tend to form regular polyhedrons (icosahedron, dodecahedron, etc.). Moreover, at these dimensions, most of the atoms that form nanoparticles are at the surface. In these conditions, it is obvious that the notion of superficial tension no longer makes any sense. Other arguments must be put forward, as we will see later.

2.2.6 Phase diagrams of nanoparticles

If fusion temperature is a function of the size of nanoparticles, it is obvious that it must be the same for the phase diagrams of non-monatomic materials. Regarding binary systems, the Gibbs free energy of a mechanical mixture is given by

$$g_m = x_1 h_1 + x_2 h_2 - T(x_1 s_1 + x_2 s_2) \tag{2.22}$$

where x_1 and x_2 are the atomic fractions of elements 1 and 2; h_i and s_i are the corresponding enthalpy and entropies. The system's entropy increases because of the configuration of the mixture

$$\Delta s_m = -k(x_1 \ln x_1 + x_2 \ln x_2) \tag{2.23}$$

When the interactions between first and second atoms are the same as in pure compounds, the solution is said to be ideal, and Gibbs energy is

$$g_{id} = g_m - T\Delta s_m = x_1 \mu_1 + x_2 \mu_2 \tag{2.24a}$$

$$\mu_i = h_i - Ts_i - kT \ln x_i \tag{2.24b}$$

Given that the mixture's entropy depends only on the total number of atoms in the system, the presence of the surface modifies Δs_m. Only the h_i are modified by the surface, via the superficial tensions γ_i. When there is no superficial segregation, the particle's energy is given by

$$g_{part} = g_{id} + x_1 g_{surf,1} + x_2 g_{surf,2} = x_1 \mu_{part,1} + x_2 \mu_{part,2} \tag{2.25a}$$

$$\mu_{part,i} = \mu_i + g_{surf,i} \tag{2.25b}$$

Let N be the total number of atoms in the particle, and x and (1–x) respectively the relative proportions of atoms 1 and 2. Then:

$$N g_{part} = x(N\mu_1 + fN^{2/3}\gamma_1) + (1 - x)(N\mu_2 + fN^{2/3}\gamma_2)$$

$$= N g_{id} + fN^{2/3}\Gamma(x) \tag{2.26a}$$

$$\Gamma(x) = x\gamma_1 + (1 - x)\gamma_2 \tag{2.26b}$$

We can, as shown previously, establish the equilibrium conditions between the liquid and solid phases:

$$N(g_{\text{part,s}} - G_{\text{part,L}}) = N(g_{\text{id,s}} - g_{\text{id,L}}) + fN^{2/3}(\Gamma_s(x) - \Gamma_L(x)) \qquad (2.27)$$

where the suffixes s and L are related to the solid and liquid phase.

In binary systems, the liquid–solid transition is defined by the solidus/liquidus curves. Solidus/liquidus curves of ideal solutions are calculated from the simultaneous equations obtained by expressing the chemical potential equality of the two phases.

$$kT \ln(x_{\text{solidus}}/x_{\text{liquidus}}) = C_1(1 - T/T_{m,1}) \qquad (2.28a)$$

$$kT \ln[(1 - x_{\text{solidus}})/(1 - x_{\text{liquidus}})] = C_2(1 - T/T_{m,2}) \qquad (2.28b)$$

where x_{solidus} and x_{liquidus} define the solidus and liquidus curves at T constant. $T_{m,1}$ and $T_{m,2}$ are respectively the fusion temperatures of the elements 1 and 2.

When the size of the particles varies, $T_{m,1}$ and $T_{m,2}$ vary; this is described in (2.18). There is no indication that the latent heat of fusion varies with R. We can then combine (2.18) and (2.28) to evaluate the solidus/liquidus curves of nanoparticles. The example of the ideal solution Ge-Si is shown in Figure 2.5.

We notice that the general shape of the solidus/liquidus curves remains the same when the size decreases. Nevertheless, for a given x concentration, the decrease

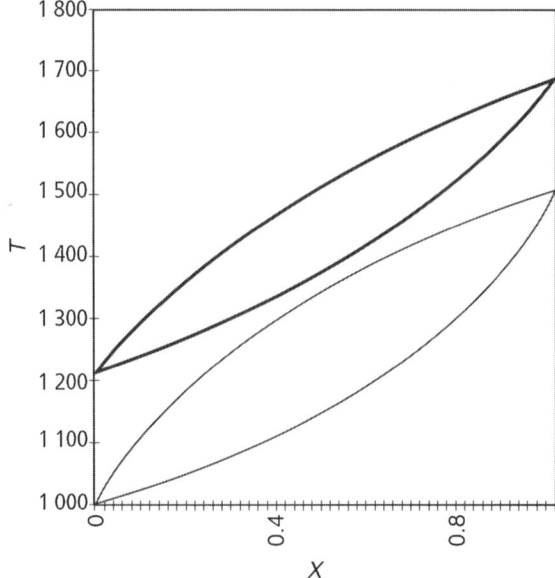

Figure 2.5   Phase diagram. Solidus/liquidus curves in the phase diagram of an ideal solution $Ge_{1-x}Si_x$. x is the atomic concentration. The temperature T is in K. Upper curves (in bold): bulk material. Lower curves: small particle containing 10^6 atoms. [Source: M. Wautelet et al., Nanotechnology, Vol. 11 (2006), 6.]

of the liquidus and solidus curves is different. For instance, in the case described in Figure 2.5, at $x = 0.5$, the liquidus curve decreases from 1 510 K for the volume to 1 345 K for the nanoparticle (a decrease of 11 per cent), while the solidus curve decreases from 1 375 K to 1 145 K (a decrease of 17 per cent). Hence the solidus/liquidus curves are distorted. Various theoretical models that predict the various distortions of the solidus/liquidus curves exist. Moreover, at fixed temperatures between the highest melting point of the bulk material and the lowest melting point of the nanoparticle, the phase diagram of the particle differs from the bulk material. This means that the relative concentration of the solid and liquid phases is different in the nanoparticle and bulk material.

Another effect modifies the behaviour of the phase diagram with the size of the nanoparticle: surface segregation. Surface segregation exists when the stoichiometry of the surface differs from that of the bulk material. In some nanoparticles (for instance, Ni-Al), a behaviour different from that of bulk alloys has been found. This is due to the fact that the structural defaults, due to variations of the stoichiometry, gather at the surface. The structures are not perfect crystals, but the nanoparticles appear to be composed of a very well-ordered centre, surrounded by a disorganized mantle, in which the deviations of the ideal stoichiometry concentrate. The surface segregation leads to a new distribution of atoms in the heart and at the surface.

Let $A_x B$ be our binary system, with N atoms; $Nx/(1+x)$ are A atoms and $N/(1+x)$ are B atoms. Assuming that the geometrical shape remains constant, the number of atoms at the nanoparticle's surface is

$$N_s = fN^{2/3} \tag{2.29}$$

The composition at the surface is $A_{xs}B$. Hence, the number of atoms in the heart is

$$N_b = N - N_s = N - fN^{2/3} \tag{2.30}$$

The composition of the heart is $A_{xb}B$.

The effect of the surface segregation is introduced by the number of atoms A and B, and, by assuming that the surface is composed of a unique atomic layer, we obtain

$$2S.x_b = -(1 + S - R) + [(1 + S - R)^2 + 4Sx]^{1/2} \tag{2.31a}$$

$$R = S(1 + x) + fN^{-1/3}(1 - S)(1 + x) \tag{2.31b}$$

From these equations, we deduce that x_b (the heart stoichiometry) and x_s (the surface stoichiometry) depend on N and x, at T and E_{segr} fixed. The compositions also depend on the geometric shape of the particle, via the term f.

An example of the effect of surface segregation on the phase diagram of ideal solutions is shown in Figure 2.6. We can see that, compared with the bulk material, the solidus/liquidus curves of the heart and the surface are shifted. It is important to note that despite the different shapes of the solidus/liquidus curves for the heart and the surface, there is no difference between their melting-point temperature but a distinct variation of composition.

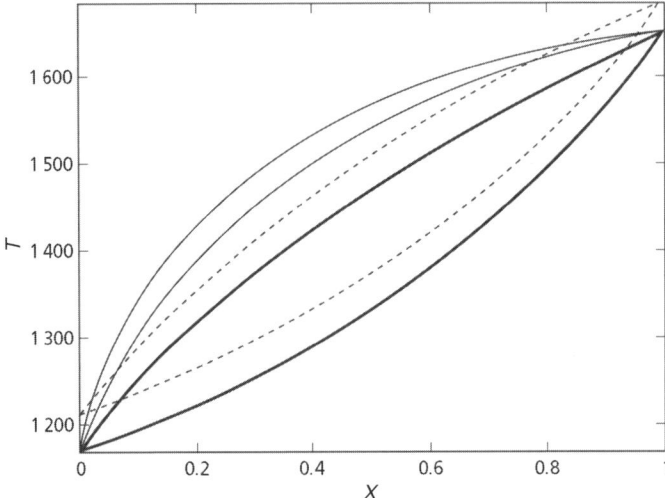

Figure 2.6   Phase diagram. Solidus/liquidus curves for the system $Ge_{1-x}Si_x$ in both
macroscopic system (dash curve) and a nanoparticular system including
106 atoms (solid curve). In this ideal model, which includes segre-
gation, the bulk (thick curve) and surface (thin curve) compositions
are considered in the nanoparticular case. [Source: R. Vallée et al.,
Nanotechnology, Vol. 12 (2001), 68.]

2.2.6.1   Effects of stresses

In most nanotechnologies, particles are not isolated but linked to a matrix (nonorganic, for instance) or are in close contact with another material. This is the case
particularly in nanoelectronics, where quantum wells, which are responsible for electronics properties, grow inside a semiconducting matrix by epitaxial growth. This is
why we can observe nanoclusters of In(Sb, As) in InAs, of (In, Ga)Sb in GaSb, of
(Cd, Zn, Mn)Se in (Zn, Mn)Se, etc. Often, those nanoclusters are compressed in the
matrix, particularly because of mesh differences between the nanoclusters and the
matrix.

When particles are compressed in a given environment, a new term must be
added in the expression of the free energy of the particle. This additional term can be
expressed as

$$G_{pr} = B.\delta V \tag{2.32}$$

where δV is the change in the atomic volume due to stresses; B is a constant proportional to the compressibility of the material. By introducing this term in the expression
of the total energy of the nanoparticle, and going back to the reasoning of the previous
paragraphs, we deduce that the nanoparticle's phase diagram changes from that of
the free particle.

When we also take into account the possible orientations of the interfaces,
we understand that the problem cannot be solved in a unique way. We note the

coexistence, in the same matrix, of domains of atomic structure of different size and orientation. Moreover, differences of free energy between the various possible structures can happen via relatively mild thermal processing. New phases can even appear. The knowledge of these phases is essential because the atomic arrangement imposes the physicochemical properties (particularly electronic properties) of nanosystems.

2.2.7   Nanoparticle stability

One of the important parameters in determining the effects of the size of the nanoparticle's phase diagram is interfacial tension. When the nanoparticle is covered by a layer of different nature, the interfacial tension is different from that of the free nanoparticle. The result is that the phase diagram is modified. One of the consequences is that it becomes possible to stabilize a chosen crystalline phase rather than another one only by covering it with a layer of a certain composition.

We can see that the issue of nanoparticle cohesion is not a simple transposition of the common matter's case. Different mechanisms affect the phase diagram of the nanoparticle: the size, exact shape, stresses, the chemical environment are critical factors.

During the nanoparticle's synthesis, many stages often happen in thermodynamic non-equilibrium conditions. The way the non-equilibrium states evolve into thermodynamic equilibrium state also affects the nanosystem's stability in the long term.

2.3   From the atom to the nanoparticle

The thermodynamic approach is applicable when the number of atoms in the nanoparticle is high enough, i.e. superior to a few thousands. Below this number, the concept of superficial tension is not valid. In order to understand the properties of these systems, it is useful to take a different approach from the one described in the previous paragraphs, i.e. to start from the atom going to the molecule, then the cluster and the nanoparticle: the *bottom-up* approach.

2.3.1   Atom clusters

When several atoms are gathered, a molecule is obtained. From which dimension can we tell that we deal with a molecule or a nanoparticle? In the following, we consider assemblages constituted of only one type of atom. Moreover, we assume that the interatomic interactions are expressed by a Lennard-Jones-type potential:

$$V_{LJ}(r) = -2/r^6 + 1/r^{12} \tag{2.33}$$

where r is the interatomic distance.

When the atoms' number N increases, the geometrical arrangement of the atoms varies. The atoms gather in clusters. Some examples are illustrated in Figure 2.7.

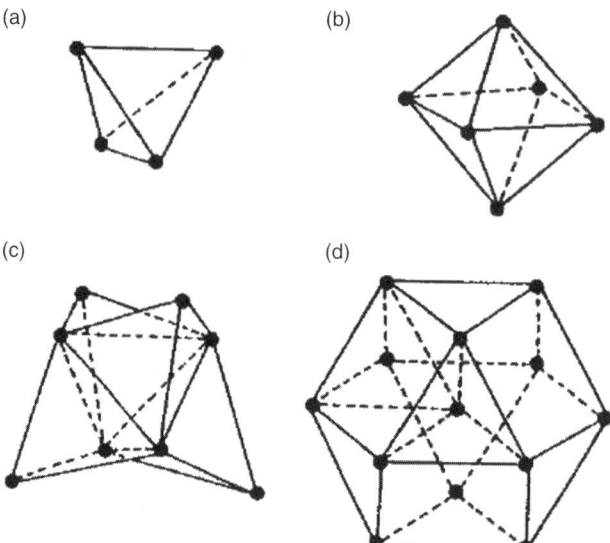

Figure 2.7   Atomic structures of some clusters: (a) $N = 4$: tetrahedron; (b) $N = 6$: hexahedron; (c) $N = 8$: twinned tetrahedron; (d) $N = 13$: cuboctahedron

When N is high enough, many geometrical arrangements are possible for the same value of N. For instance, when $N = 6$, two local minima of the total energy appear. They correspond to a stable structure (the octahedron), and a metastable one (the tripyramid).

When $N = 7$, the number of local minima is four, corresponding to the structures illustrated in Figure 2.8. This number of local minima increases rapidly with N, as shown in Table 2.3.

The different local minima are characterized by energies situated in a relatively low region. Many metastable states show energies that are close to the one at the fundamental level. If these energy differences are comparable to the thermal energy, kT (T is the temperature of the system, k is the Boltzmann constant), we can suppose that there will be thermal fluctuations of the atomic structure. The presence of many metastable states at the same energy (or almost the same) is also an indication that the excitation energy can be dissipated in many states, and be stored for quite a long time. This points out as well, as Berry demonstrated in 1990, that the fusion temperature and the solidification one for small particles can be different.

Those interpretations are confirmed by the theory. The total energy variation of those particles with temperature is calculated by molecular dynamic methods. In the case of clusters with N as small as 13, for instance, it has been demonstrated that fusion temperatures and solidification temperatures are different. At low temperatures, the cluster is 'solid': the positions of the atoms do not change very much. At high temperature, the cluster is 'liquid': the average length of interatomic bonds

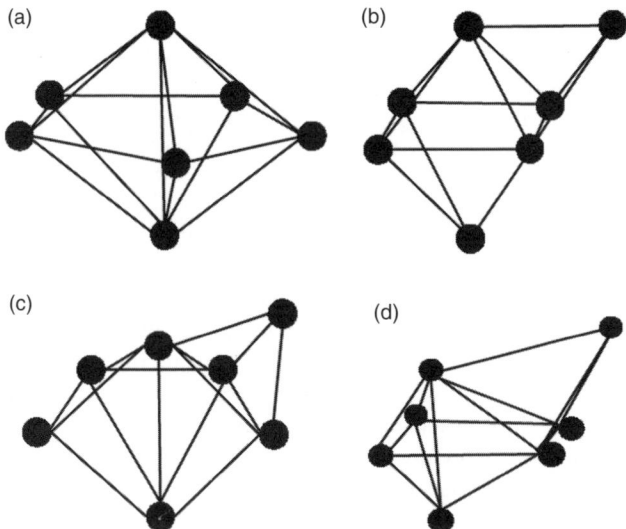

*Figure 2.8   Stable and metastable atomic structures of clusters of seven atoms: (a)
pentagonal bipyramid; (b) octahedron plus one; (c) twisted structure,
(d) tetrahedron group*

*Table 2.3   Number of local minima of the total energy
of a particle of N atoms*

N	6	7	8	9	10	11	12	13
	2	4	8	18	57	145	366	988

fluctuates a lot around the equilibrium states. Regarding intermediate temperatures,
the 'solid' and 'liquid' states coexist. When we start from a low temperature and we
heat the system, the fusion occurs at a given temperature. However, if we start from
a high temperature where the cluster is melted, and we cool down the system, the
solidification happens at a different temperature from the fusion one. It is important
to note that these results show that it is possible to define 'solid' and 'liquid' states
for those very small particles.

In those clusters, the fusion temperature is lower than in bulk materials, which
confirms the observed trend in nanoparticles. However, when the size increases up
to $N \approx 100$, another phenomenon appears. A German research team shows, on measurements of sodium seeds that contain 70 to 200 atoms, that the fusion temperature
is not fixed; it seems to vary with the grain shape. Moreover, the researchers showed
that the latent heat of fusion varies with the size of the grain. Consequently, they concluded that it is at this size that fusion is the result of many interdependent parameters,
such as size, atomic arrangement or even electronic structure.

2.3.2   Nanoparticles

When the nanoparticle size increases with N reaching a few hundreds, the computation of atomic properties becomes very difficult because of the required computation times. Experiments show that interesting phenomena still happen.

2.3.3   Magic numbers

A first phenomenon was noticed in the 1980s. A beam of sodium particles is produced by adiabatic expansion of a gas composed of evaporated argon and sodium. After ionization of the sodium particles, the analysis of the mass spectrograph shows that there are some anomalies of abundance for values of $N = 8, 20, 40, 58, 92$. These numbers are known as *magic numbers*. A magic number corresponds to a specific size of cluster where we find an abundance anomaly in the mass spectrum. This is an indication that clusters of these sizes are stable compared with those with our size. It has been theoretically demonstrated that magic numbers are associated with the shell-closing numbers of valence electrons of Na: the valence electrons move independently in an effective potential with spherical symmetry. It is interesting to note that this issue is similar to the one of the shell structure of atomic nuclei.

Magic numbers are also detected in particles of bivalent metals (Cd, Zn) and trivalent (Al), as well as in ions of transition metals (Cu, Ag, Au).

2.3.4   Fullerenes

Magic numbers also appear in the case of carbon, but their interpretation is different. Moving on to consider the importance of carbon particles and their derivatives, let us take a closer look at this case.

In 1985, Kroto *et al.* vaporized graphite by laser beam in a chamber full of helium; they then measured the mass spectrum of carbon. They found that particles with 60 and 70 atoms were exceptionally stable. After some experimentation, they came to the conclusion that the particle has a cage shape, made of an assembly of pentagons and hexagons, as illustrated in Figure 2.9.

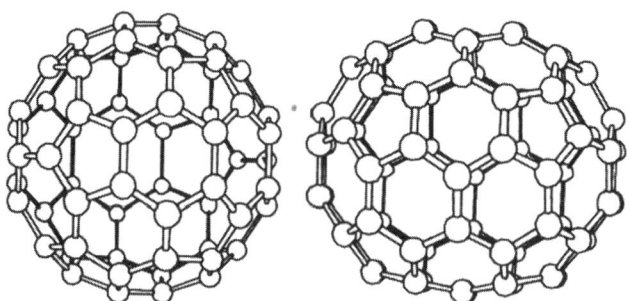

Figure 2.9   Structures of C60 and C70

This structure has been named a *fullerene*, from the name of architect R. Buckminster Fuller, who built domes composed of pentagons and hexagons. Since Kroto, methods allowing the production of usable and tradable amounts of fullerenes have been developed.

Fullerenes form a new allotropic form of carbon, like graphite and diamond. In graphite, outer electrons form a sp^2 hybrid orbital. The carbon atoms are arranged in planes composed of hexagons. In diamond, the outer electrons form a sp^3 hybrid orbital. The carbon atoms are in the centre of tetrahedrons composed of other identical carbon atoms. In fullerenes, they are in mixed hybridization because of the curvature of their structure. It is the ratio between the number of pentagons (always 12) and the number of hexagons that gives the properties of the various fullerenes.

Geometry of fullerenes

Fullerenes have the shape of convex polyhedrons. A convex polyhedron, with a spherical topology, can be characterized by the polygons that cover its surface. These polygons, which are F in number, have S vertices and A edges in total. There's a relation between these three numbers, called Euler's relation:

$$S - A + F = 2 \qquad (G.1)$$

The number 2 corresponds to a sphere topology. This number is equal to 0 for a torus. If we want to build a polyhedron with H hexagons and P pentagons, then

$$F = H + P \qquad (G.2)$$

$$2A = 6H + 5P \qquad (G.3)$$

$$3S = 6H + 5P \qquad (G.4)$$

The factor 2 in (G.3) comes from the fact that a vertex is common to two polygons. In the same way, the factor 3 in (G.4) comes from the fact that a vertex is common to three polygons.

By transferring those equations in (G.1), we get

$$P = 12 \qquad (G.5)$$

The value of H depends on the conditions given for S.

In the fullerene C_{60}, the 60 atoms are identical. It is not the case in the C_{70}, where atoms are divided into five groups according to the type of bonds with their neighbour regarding 1:1:1:2:2. C_{70} has the shape of a rugby ball. Regarding the other C_N, stable shapes exist for $N = 180, 240,\ldots, 540,\ldots$.

Fullerenes being hollow atomic assemblies, foreign atoms can be introduced into them, particularly metallic atoms (La, Ni, Na, K, Rb, Cs). Fullerene molecules can also be arranged in a crystal lattice.

C_{60} is a particular molecule: almost spherical with respect to a five-order axis. It is therefore impossible to pave space with such molecules. At ambient temperature, C_{60} molecules form a face centred cubic structure. In this structure, C_{60} molecules seem to turn freely, independently from the others. When the temperature decreases, a phase shift to a simple cubic structure occurs around 250 K.

In the crystalline structure, we can also introduce foreign atoms (K, Rb,...). Some of the compounds obtained in this way are supraconducting at low temperature. So, the Curie temperature of K_3C_{60} is 19 K, that of Rb_3C_{60} is 28 K, and that of Cs_2RbC_{60} is 33 K.

2.3.5   *Nanotubes*

Carbon can also take a nanotube form. Carbon nanotubes can be considered as sheets of a few atoms wide rolled up on themselves and closed at the two ends. The diameter of those nanotubes is from one to a few nanometres. The length can reach 10 to 100 μm.

Nanotubes are cylindrical structures based on graphite hexagonal lattice. Three types of nanotube are identified, depending on how the two-dimensional carbon sheets are rolled up. These nanotubes are known as *armchair*, *zigzag*, and *chiral* (see Figure 2.10). Multi-walled nanotubes can also be found; they consist of multiple layers of graphite rolled in on themselves.

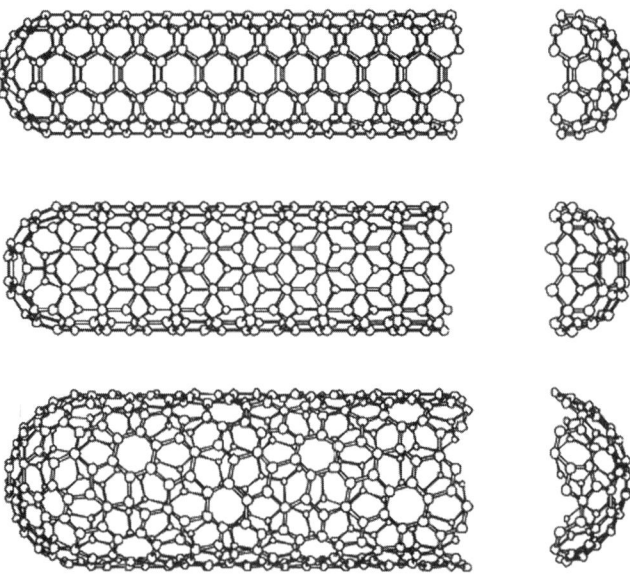

Figure 2.10   *By rolling up a graphene sheet (a monolayer of a graphite crystal), and by closing each end by half a C_{60} molecule, we obtain nanotubes of three different possible shapes*

Carbon nanotube structures

Carbon nanotubes result from the rolling-up of a monolayer of graphite, known also as a graphene sheet. Two quantities define the nanotube's geometry: the diameter d_N, and the chiral angle θ. The nanotube circumference is expressed as a function of the chiral vector, $C_h = n\,a_1 + m\,a_2$, which links two equivalent crystallographic regions of a graphene sheet (Figure 2.11).

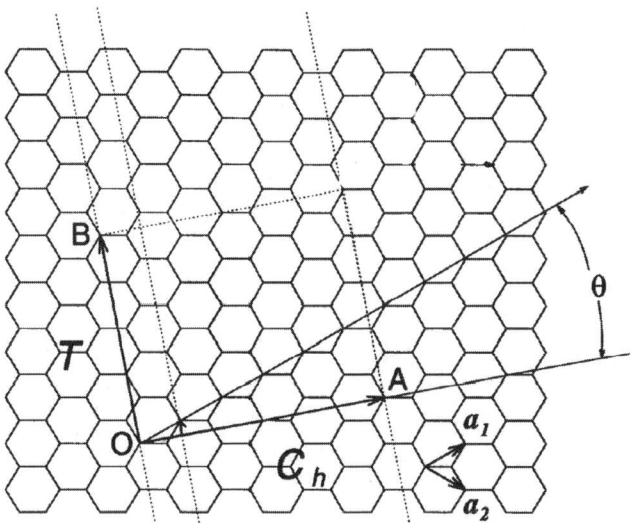

Figure 2.11   *Elements defining the nanotube's nanostructure on a graphene sheet*

The chiral angle θ is the angle between the chiral vector and a_1 vector of the hexagonal graphene lattice. The chiral vector is named after the couple of integer numbers (n, m). The intersection of OB with the first point of the graphene lattice fixes the fundamental translation one-dimensional vector of the nanotube. The cylinder that defines the nanotube is obtained by joining the two ends of C_h vector. Angles of interatomic bonds are slightly distorted compared with the graphene because of the nanotube curvature. In the (n, m) notation, $(n, 0)$ and $(0, m)$ vectors give a *zigzag* nanotube, (n, n) vectors give an *armchair* structure, and the other (n, m) vectors give a *chiral* structure. The nanotube diameter is given by:

$$d_N = a_C \sqrt{3}(m^2 + mn + n^2)^{1/2}/\pi$$

and the chiral angle is defined by:

$$\theta = \arctan(\sqrt{3}n/(2m + n))$$

Because of the outstanding electronic and structural properties of nanotubes, a lot of research work has been done in this area. Nanotubes can be metallic or semi-conducting, depending on the number of hexagons on the circumference and the orientation of the hexagons regarding the nanotube axis. They also have excellent mechanical resistance.

2.3.6   Filling of nanotubes

Carbon nanotubes are hollow. Right after they were discovered, the idea of filling them with different materials became immediately apparent. In this way, nanotubes can be used as nano-test-tubes or nanotanks. On the other hand, due to its large specific surface, the outer surface could be covered by a material to carry, or by a catalyst.

For such applications it is necessary to understand the wetting properties of nanotubes. These properties define the ability of a liquid to spontaneously spread on the nanotube's outer surface or to get in by capillary action etc. Capillarity is an indicator of wetting properties. This is described in the Young–Laplace equation:

$$\Delta p = (2\gamma/r)\cos\theta \tag{2.35}$$

In this relation, Δp is the difference pressure at the interface liquid–vapour in the capillary. γ is the liquid superficial tension and θ is the contact angle between the meniscus and the liquid. The curvature radius of the meniscus is r.

The liquid is directly sent into the capillary when $\Delta p > 0$, that is to say $\theta < \pi/2$. Actually, it is difficult to theoretically predict whether a liquid will get into the nanotube or not. Experimentally, we note that liquids with a low superficial tension get into carbon nanotubes. Some examples are summarized in Table 2.4.

Table 2.4   Nanotube wetting properties of some liquid matter [Source: T.W. Ebbesen, Physics Today, June 1996, p. 31]

Matter	Superficial tension (mN/m)	...gets into nanotube?
HNO_3	43	Yes
S	61	Yes
Cs	67	Yes
Rb	77	Yes
V_2O_5	80	Yes
Se	97	Yes
Te	190	No
Pb	470	No
Hg	490	No
Ga	710	No

2.3.7 Geometric shape of non-hollow clusters

Let us go back to a non-hollow assembly of atoms. When the size increases to reach a few hundreds of atoms, regular geometrical shapes appear. The most frequent shapes, which are also the most interesting, are illustrated in Figure 2.12.

The presented geometrical shapes are the cuboctahedron, the icosahedron, the regular decahedron, the star decahedron, the Marks truncated decahedron and the round decahedron. The first shape, generic of the face-centred cubic structure, has many variants. Decahedrons are derivatives of the same shape. They correspond to one of the most oft-found shapes, because of their high stability. These geometric shapes can be found in many metals and in silicon.

The icosahedron shape is found in particles of dimensions between 1 and 100 nm. During their analysis by electronic microscopy, icosahedrons can be differently aligned, depending on the experimental conditions. Therefore, image analyses are needed for a complete identification of particles and their orientations.

As in the case of the smallest particles, the relative stability of different shapes is important. However, given the high number of atoms, the analysis of total energies

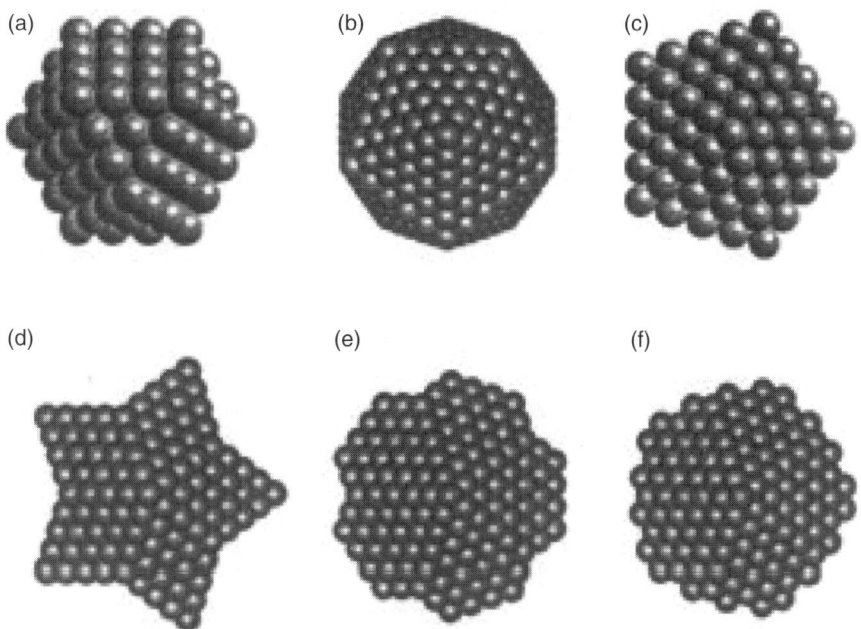

Figure 2.12   Most common cluster shapes: (a) cuboctahedron; (b) icosahedron; (c) regular decahedron; (d) star decahedron; (e) Marks decahedron; (f) round decahedron [Source: Yacaman et al., 2001, American Institute of Physics]

corresponding to the different forms is long. Such analyses have been conducted by Landman's and Yacaman's teams.

The most stable structures correspond to the truncated decahedrons: star, rounded and Marks decahedrons. For the smallest particles, the icosahedron and the regular decahedron are more stable than the face-centred cubic structure. When the size increases, the truncated decahedron structures remain stable while the icosahedron and the regular decahedron are less favoured.

Furthermore, fabrication methods of nanoparticles are such that their growing usually happens outside thermodynamic equilibrium conditions. Consequently, we note a diversified distribution of nanoparticle shapes. This is the case during vapour-phase growing. However, with colloidal methods (slower methods), one form can be better than others.

Similarly with big particles, internal tensions are present when nanoparticles are growing. Relief mechanisms for tensions must also occur during the establishing of the nanoparticles' exact shapes. In the decahedron particle's case, elastic models show that many mechanisms are possible: (a) dislocation formation; (b) formation of a system of thin, twinned, parallel layers in one section of the decahedron; (c) separation of the pentagonal axis in two (or more) partial declinations; (d) displacement of the pentagonal axis to the particle's periphery. These different mechanisms have been observed experimentally.

2.3.8   Shape fluctuations

When nanoparticles are heated and/or are subject to the action of an electron beam (in an electronic microscope, for instance), we can see continuous variations in shape. In 1987, two Japanese researchers, S. Iijima and T. Ichihashi conducted an experiment that has been the starting point of numerous researches on this subject. They analysed, through an electronic microscope, a gold grain of around 2 nm in diameter, and watched its behaviour via video recording, with a temporal resolution of 1/60 s. The grain was composed of around 460 atoms. Different shapes were detected at different times. Cuboctahedrons were observed, which resulted from a face-centred cubic structure. There were also icosahedrons, which are never found as local structures of periodic crystals. Spherical grains that look like liquid drops could also be detected.

Therefore, it is clear that, at this scale, the grain's structure changes continuously. The grain behaves like neither a solid nor a liquid. Other researchers have since noted similar behaviours for various metals. The fluctuation between different structures seems to be a phenomenon inherent to the microscopic behaviour of small particles.

Many interpretations of this phenomenon are proposed. For some of them, nanoparticles completely melt due to the heat resulting from the interactions with the electrons, then re-crystallise into various structures. For the others, given the low total energy gap between the different forms and the low activation energies, the fluctuation is easy between the different forms. This phenomenon has been named *quasi-fusion*. Currently it is not possible to favour one of the two interpretations.

As we can see, nanoparticle cohesion is a huge field of research. This cohesion is essential because the physical and chemical properties of nanosystems depend on the exact shape of particles and on the atoms' arrangement.

Chapter 3
Electronic structures of nanosystems

Particles that vary in size between 1 and 100 nm have specific physicochemical properties. These nanoparticles are in an intermediate state between the solid and the molecule. In a crystalline solid, electrons occupy states in continuous energy bands. The analysis of these energy bands becomes easy due to the periodical arrangements of the atoms. This field is known as *solid-state physics*. The width, the division, the occupation of the electronic energy bands gives the electrical, optical and magnetic properties of solids.

At the other end of the dimensions scale, the electronic-state density of atoms and molecules is discontinuous. The electronic energy levels are calculated by well-known methods of atomic and molecular physics, and quantum chemistry. These analyses require powerful computers because the calculi are heavy and complex. And, the higher the number of atoms, the more powerful the computers must be. To analyse the properties of nanoparticles composed of a few hundred atoms is almost impossible.

When looking at nanosystems, analysis simplification of the two extremes described above is no longer possible. We have to introduce the concept of an ideal nanocrystal, known as a *quantum dot*, which will help explain the particular properties of nanosystems.

First, we will summarize some essential elementary knowledge of electrons' behaviour in the matter.

3.1 Electrons in matter

The behaviour of electrons in matter is governed by the laws of quantum mechanics.

3.1.1 *An electron in a one-dimensional potential well*

Let us start with the simplest case: an electron moving in a one-dimensional potential. An electron of mass m is enclosed in a segment of length a restricted by infinite barriers. The electron's behaviour is described by its wave function $\psi_n(x)$, a solution of the time-independent Schrödinger equation:

$$H\psi_n(x) = [-(\hbar^2/2m)\partial^2/\partial x^2 + U(x)]\psi_n(x) = E_n\psi_n(x) \tag{3.1}$$

H is the Hamilton operator of the system. $U(x)$ is the potential seen by the electron of mass m. E_n are the energy eigenvalues of the electron, depending on the shape and the depth of $U(x)$. The potential $U(x)$ is described by:

$$U(x) = 0, \text{ for } |x| \leq a/2; \quad U(x) = \infty, \text{ for } |x| > a/2 \tag{3.2}$$

In these equations, a is the width of the potential well. The electron's behaviour is described in all quantum-mechanics elementary courses. The wave functions are given by

$$\psi_n(x) = A \cos(2\pi x/\lambda_n) \tag{3.3a}$$
$$\lambda_n = 2a/n \tag{3.3b}$$

The electrons occupy discontinuous energy levels, given by

$$E_n = (\pi^2 \hbar^2/2ma^2)n^2 \tag{3.4}$$

The first energy levels and their associated wave functions are illustrated in Figure 3.1.

The separation between the adjacent energy levels increases with the quantum number n:

$$\Delta E_n = E_{n+1} - E_n = (\pi^2 \hbar^2/2ma^2)(2n + 1) \tag{3.5}$$

Because $U(x) = 0$ in the potential well, (3.4) gives the value of the electron kinetic energy, the movement quantity p, and the wave number k:

$$E = p^2/2m; \quad p = \hbar k \tag{3.6}$$

We can deduce that p and k have discrete values.

When the depth of the potential well has finite value U_0, the wave functions do not cancel out around the well but exponentially decrease in the classically forbidden region. Given the quantum meaning of the wave function, it means that the electron has an above-zero probability of being outside the potential region of the well. This probability increases with n. The number n of energy levels in the potential well is given by the relation

$$a(2mU_0)^{1/2} > \pi \hbar(n - 1) \tag{3.7}$$

The absolute position of the energy levels is a little lower when U_0 is finite compared with the case where U_0 is infinite, because the electron's effective wavelength is a wider. The states characterized by $E_n > U_0$ are associated with an infinite movement and form the state continuum.

As an illustration, let us consider the case of a free electron being in an infinite potential well of width $a = 1$ nm. We calculate $E_1 = 0.094$ eV; $E_2 = 0.376$ eV; .... To compare, remember that, at ambient temperature, $kT = 0.025$ eV; if the transition from E_1 to E_2 is done by photon absorption, the corresponding wavelength is equal to $\lambda = 4\,394$ nm, in the middle of the infrared range.

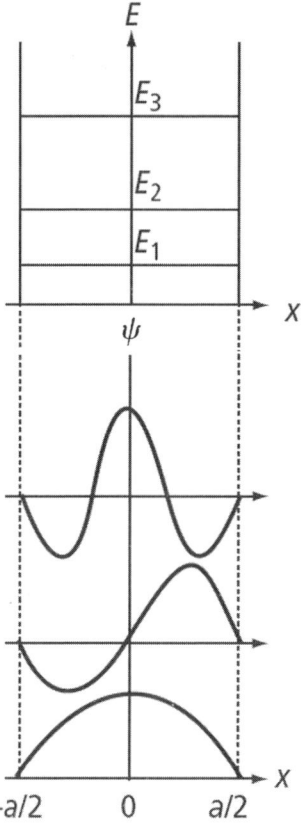

Figure 3.1   *Free electron in a box with infinitely high potential walls. Three first energy levels and wave functions of a free electron in a box with infinite walls, of width* a. *The energy levels are numbered after the quantum number* n

3.1.2   An electron in a spherical potential well

The Hamiltonian H describing the behaviour of a particle in a spherically symmetric potential is

$$H = [-(\hbar^2/2m)\Delta + U(r)] \tag{3.8}$$

where $r = \sqrt{(x^2 + y^2 + z^2)}$. The simplest case is the one of a spherical potential well with an infinite barrier:

$$U(r) = 0, \text{ for } r \leq a; \quad U(r) = \infty, \text{ for } r > a \tag{3.9}$$

In this case the energy levels are given by

$$E_{nl} = \hbar^2 \chi_{nl}^2 / 2ma^2 \tag{3.10}$$

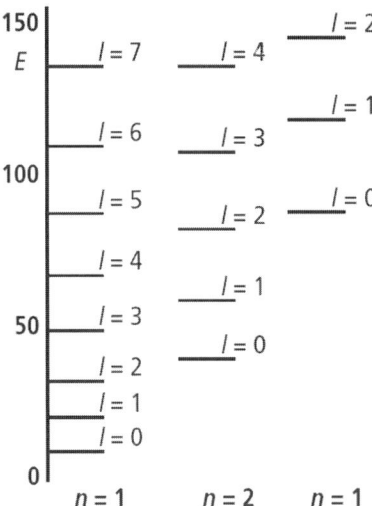

Figure 3.2   Energy levels of a particle in a spherically symmetric potential well, with infinite barriers. States are ordered as a function of the principal quantum number n and the orbital quantum number l. Each state is degenerated (2l + 1) times

In this equation χ_{nl} are the roots of spherical Bessel functions, with n being the root number, and l the function order. When $l = 0$, these values are equal to πn, and the values of the energy correspond to that of the case where the potential barrier has one dimension.

When the depth of the potential well has a finite value U_0, the energy eigenvalues are a bit different from the case of infinite potential, as far as

$$U_0 \gg \hbar^2/8ma^2 \tag{3.11}$$

When

$$U_0 = \pi^2\hbar^2/8ma^2 \tag{3.12}$$

only one state exists in the well: $E_1 = U_0$. When $U_0 < U_{0\,min}$, there is no state in the potential well. This is a main difference between the three- and one-dimensional cases.

3.1.3   An electron in a hydrogen atom

The hydrogen atom is the simplest real quantum system. An electron of mass m moves around a proton of mass M_p. The Schrödinger equation of the hydrogen atom:

$$H = -(\hbar^2/2M_p)\Delta_p - (\hbar^2/2m)\Delta_e - e^2/|r_p - r_e| \tag{3.13}$$

In this equation r_p and r_e are respectively the position vectors of the proton and the electron. The electron's energy levels, analysed in all elementary quantum mechanics

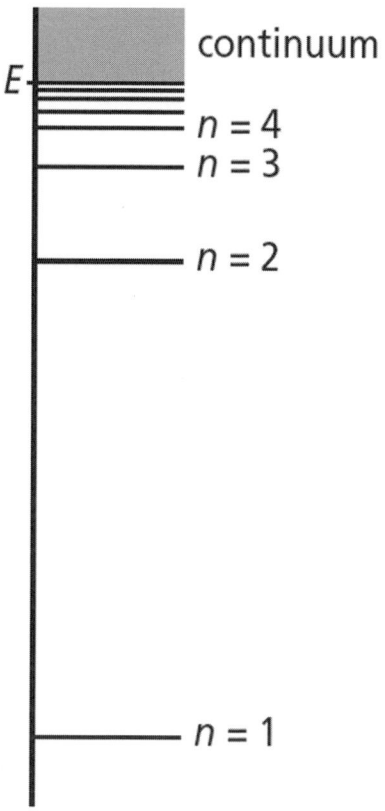

Figure 3.3   Energy levels of the electron in the hydrogen atom

books, are given by

$$E_n = -\text{Ry}/n^2 \tag{3.14}$$

where

$$\text{Ry} = e^2/2a_B; \quad a_B = \hbar^2/\mu e^2 \tag{3.15}$$

The Rydberg constant Ry corresponds to the ionization energy of the hydrogen atom's fundamental state. a_B is the Bohr radius of the hydrogen atom. μ is the reduced mass of the proton-electron system:

$$\mu = mM_p/(m + M_p) \tag{3.16}$$

The ratio $\mu/m = 0.9995$ in the case of the hydrogen atom.

The expression of the energy is valid in other systems where a particle moves around another one of opposite charge. In this case, the reduced mass can be very different from that of hydrogen, as we will see in the case of the exciton.

3.1.4   An electron in a periodical potential

When moving in a crystal, an electron perceives a periodic potential. Consider the case of a one-dimensional system. The potential seen by the electron is described by:

$$U(x) = U(x + a) \tag{3.17}$$

This potential is invariant during every translation of distance a. As it is demonstrated in all solid-state books, the behaviour of electrons is described by a wave function $\psi(x)$ that satisfies Bloch's theorem:

$$\Psi(x) = \exp(ikx)u_k(x); \quad u_k(x) = u_k(x + a) \tag{3.18}$$

This equation shows that the wave function associated with a Hamiltonian with a periodic potential is a plane wave modulated by a function that has the same period as the potential. It can also be noted that the wave functions depend on the number k. In other words, the wave function and the associated energy levels depend only on k. The wave numbers k_1 and k_2 are equivalent.

$$k_1 - k_2 = (2\pi/a)n, \quad n = \pm 1, \pm 2, \pm 3 \tag{3.19}$$

This is a direct consequence of the translation symmetry of the system. Consequently, we can describe everything as being in k intervals of width $2\pi/a$. This interval defines what we call the *Brillouin zone*.

The quantity $p = \hbar k$ is called the crystalline moment or quantity of movement of the electron in the system. It differs from the quantity of movement by a particular conservation law: the quantity of movement is conserved to within about $2\pi h/a$.

The energy spectrum of the electron in a periodic potential is described by a function $E(k)$. When the potential is constant ($U(x) = 0$), the electron is said to be free, and

$$E(k) = \hbar^2 k^2 / 2m \tag{3.20}$$

When $U(x) \neq 0$, $E(k)$ differs from the case of the free electron, as we can see in Figure 3.4.

However, $E(k)$ is generally expressed as:

$$E(k) = \hbar^2 h^2 / 2m^*(k) \tag{3.21}$$

The term $m^*(k)$ is called the electron's *effective mass*. Often, we can consider that $m^*(k)$ is constant. This is true around an extreme of $E(k)$; we can write:

$$m^{*-1} = (1/\hbar^2)(d^2 E/dk^2)_{k=0} \tag{3.22}$$

This effective mass gives the reaction of the electron to an external force F:

$$F = m^* \gamma \tag{3.23}$$

where γ is the acceleration.

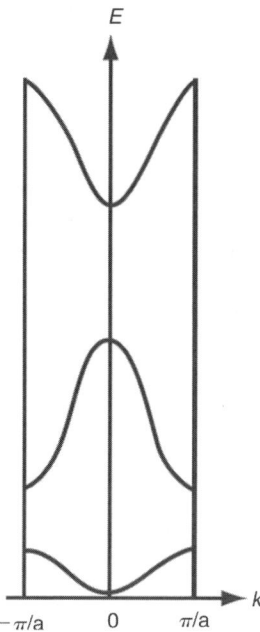

Figure 3.4   Electron in a linear lattice. Energy as a function of the wave vector k, for an electron belonging to a linear lattice, of parameter a, with $U(x) \neq 0$

By analysing the case described in Figure 3.4, we can note that, around the Brillouin region, the effective mass is lower than that of the free electron. An electron subject to a periodic potential may seem 'lighter' than a free electron. Sometimes, it can seem 'heavier'. It can even have a negative effective mass, when the electron has an energy close to a maximum of $E(k)$. This negative effective mass is not an artefact but an important property of a particle that interacts simultaneously with a periodic potential and an external force. A negative effective mass means that the particle's quantity of movement decreases under the action of an external force.

The interpretation can easily be extended to the case of a three-dimensional potential, which corresponds to the case of a crystalline solid.

As we have just seen, the electron's energy spectrum consists of energy bands separated by forbidden bands. Electronics properties of solids are determined by the way energy bands are occupied and by the absolute value of the forbidden band between the last occupied band and the first empty band. When a crystal has an empty energy band, its behaviour is metallic. If $T=0$, all bands are either full or empty, the crystal behaves like an electric insulator or a semiconductor. The last occupied band is called a *valence band*, while the first empty one is called *conduction band*. The gap between the minimum of the conduction band and the maximum of the valence band is called the *forbidden band*, E_g. Depending on the E_g absolute value, the crystal will be either insulating or semiconducting. When $E_g < 4$ eV, the crystal is a semiconductor.

3.1.5   Electron, hole and exciton

A free electron is characterized by its charge $-e$, spin $1/2$, mass m, and quantity of movement $\hbar k$. In a semiconducting or insulating crystal, an electron can be transferred from the top of the valence band to the bottom of the conduction band, for instance due to thermal agitation at $T \neq 0$. When an electron is in the conduction band, it is described by its charge $-e$, spin $1/2$, effective mass m_e^*, and quantity of movement $\hbar k_e$. Only the electric charge and the spin are conserved when dealing with a free electron.

When an electron moves from the valence band to the conduction band, it creates a free position known as a *hole*. This hole behaves like a particle (quasi-particle) characterized by its charge $+e$, spin $1/2$ (positive), effective mass m_h^*, and quantity of movement $\hbar k_h$.

By using concepts of elementary excitations, we can consider the fundamental state of a crystal as a *vacuum state*: there are neither electrons in the conduction band nor holes in the valence band. As for the first excited state, it corresponds to the formation of an electron-hole pair: an electron in the conduction band and a hole in the valence band. A transition from the fundamental state to the first excited state happens under the action of an external perturbation, like the absorption of a photon energy $h\nu$ (ν is the electromagnetic wave frequency associated with the photon; the photon quantity of movement is $\hbar k_{ph}$). In this case, there is energy and quantity of movement conservation:

$$h\nu = E_g + E_{cin,e} + E_{cin,h} \tag{3.24a}$$

$$\hbar k_{ph} = \hbar k_e + \hbar k_h \tag{3.24b}$$

In the visible domain, the photon movement quantity is negligible, and we have a transition known as vertical. The inverse transition – the annihilation of the electron-hole pair and the emission of a photon – is possible as well.

An electron-hole pair behaves like the electron and the proton of the hydrogen atom with the following differences:

- the proton is replaced by a hole, of effective mass m_h^*;
- the electron is characterized by its effective mass m_e^*;
- both move in a medium where the dielectric constant is that of the crystal, $\varepsilon \neq 1$.

An electron and a hole interact like a proton and an electron in the hydrogen atom. They form a quasi-particle that we call an *exciton*. From then on, similarly to the case of a hydrogen atom, the exciton is characterized by the exciton Bohr radius:

$$a_B^* = \varepsilon \hbar^2 / \mu e^2 = (\varepsilon m / \mu) \times 0.053 \text{ nm} \tag{3.25}$$

Table 3.1   Exciton characteristic parameters of some semiconductors

	Forbidden band E_g (eV)	Exciton's Rydberg energy Ry* (meV)	Exciton's Bohr radius a_B^* (nm)
Si	1.17	15	4.3
GaAs	1.518	5	12.5
CdSe	1.84	16	4.9
CdS	2.583	29	2.8
ZnSe	2.82	19	3.8
AgBr	2.684	16	4.2
CuBr	3.077	108	1.2
CuCl	3.395	190	0.7

μ is the reduced mass of the electron-hole pair:

$$\mu^{-1} = m_e^{*-1} + m_h^{*-1} \tag{3.26}$$

The exciton is characterized by its Rydberg energy:

$$\text{Ry}^* = e^2/2\varepsilon a_B^* = (\mu/m\varepsilon^2) \times 13.6 \text{ eV} \tag{3.27}$$

The effective mass of the electron-hole pair is lower than the mass of the electron, while the dielectric constant is greater than the vacuum one. This is why the exciton Bohr radius is slightly bigger and the Rydberg energy much smaller than the corresponding values for a hydrogen atom. In semiconductors, the exciton Bohr radius is around 1 to 10 nm, while its Rydberg energy is around 1 to 100 meV. The exciton characteristic parameters of some semiconductors are given in Table 3.1.

In a semiconductor, the forbidden band energy, E_g, corresponds to the minimal energy required to create a non-interacting electron and a hole. The energy levels corresponding to an exciton are then situated just below the forbidden band. Exciton energy levels are given by the relation ($M = m_e^* + m_h^*$)

$$E_n(k) = E_g - \text{Ry}^*/n^2 + \hbar^2 k^2/2M \tag{3.28}$$

In this relation, in the right-hand member, are included, in succession, the forbidden band, the whole energy levels equivalent to the hydrogen atom's energy, and the kinetic energy of the exciton centre of mass. Like the non-interacting electron-hole pairs, excitons can be created by the absorption of photons.

3.1.6   From zero to three dimensions

In semiconductors, the De Broglie wavelength of an electron λ_e and the hole's one λ_h, as well as the electron's Bohr radius a_B^*, can be considerably higher than the time constant of the lattice a.

Consequently, it is possible to create a mesoscopic structure for which one, two or three dimensions are comparable to or smaller than λ_e, λ_h, and a_B, but bigger than a. In these structures, elementary excitations will be subjected to a quantum confinement, of which the result will be a movement that is finite along the confinement axis and infinite in the other directions.

In the case of the confinement in one dimension, we obtain a two-dimensional structure known as a *quantum well*. When the confinement is in two directions, the resultant structure is one-dimensional; this structure is known as a *quantum wire*. When the movement of the electrons, the holes, and the excitons is restricted in the three dimensions, we say that the system is at zero dimension; this system is known as a *quantum dot.*

In structures confined in one and two dimensions, the quasi-particles (at low concentration) can be seen as an 'electron gas', as in the three-dimensional case. The state density of electrons and holes can be written according to the general form,

$$\rho(E) \neq E^{d/2-1} \tag{3.29}$$

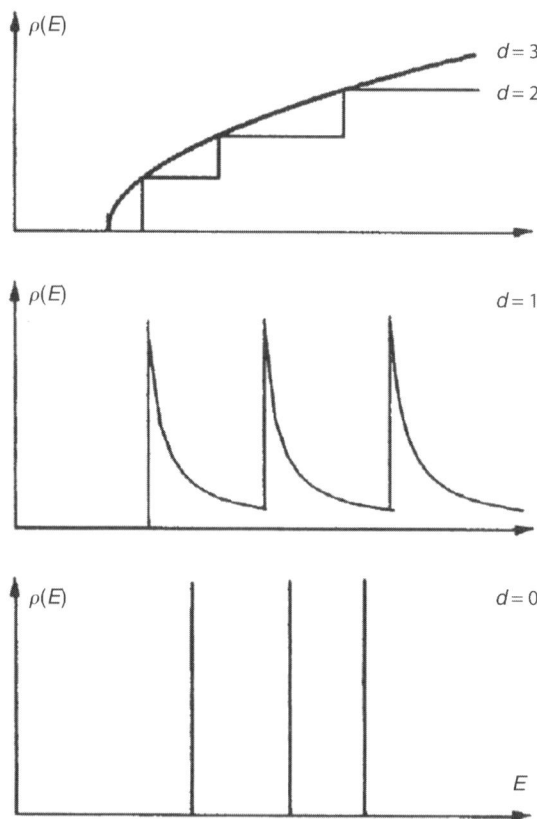

Figure 3.5   Shape of electronic state density for different dimensions

where d is the system dimension, and the energy is measured from the bottom of the conduction band for the electrons, and from the top of the valence band for the holes. In three dimensions, $\rho(E)$ is proportional to the square root of the energy. When $d = 2$ and 1, discrete sub-bands appear, because of the effect of the confinement. The state density is given by the previous equation in each sub-band. When $d = 0$, the energy levels are discrete.

The structures presented above are ideal structures. In reality, when we manufacture such structures, some inhomogeneities and/or some defects appear, which then bring a number of new properties.

3.2   From the solid to the nanoparticle

The simplest nanoparticle ideally consists of a small-sized crystal of spherical or cubical shape, known as a *quantum dot*. Although it does not exist in nature, the quantum-dot notion allows us to understand the fundamental characteristics of nanoparticles.

To go from the crystalline solid to the nanoparticle, we must first consider that the quasi-particles have the same material properties as in an infinite solid that can be modified by including effects caused by nanoparticle boundaries. Given that spatial parameters of quasi-particles (De Broglie wavelength, Bohr radius of the exciton) are greater than the dimension of the crystal lattice mesh of most semiconductors, we can consider that the nanoparticle has a great number of atoms and can be considered as a macroscopic crystal regarding the lattice properties, but as a 'quantum box' regarding the quasi-particles. We can therefore conclude that electrons and holes behave as if they were being characterized by an effective mass equal to the infinite solid one. This is called the *effective-mass approximation*.

In order to better understand the effects of quantum confinement, we are going to consider the simplest case: a box quantum confinement with spherical symmetry, the confinement containing electrons and holes of isotropic effective mass.

3.2.1   *Weak confinement*

The weak-confinement regime corresponds to the case where the radius a of the particle is small, but still a few times greater than the Bohr radius of the exciton. In this case, the centre-of-mass quantization of the exciton occurs. Going from the law of dispersion of excitons in a crystal, we replace the exciton kinetic energy by an expression extracted from the movement of a particle in a spherically symmetric potential:

$$E_{nml} = E_g - \text{Ry}^*/n^2 + \hbar^2 \chi_{ml}^2/2Ma^2 \qquad (3.30)$$

The exciton is characterized by the quantum number n describing the states due to the Coulombian interaction electron-hole (1S ; 2S 2P; 3S, 3P, 3D; ...), and by two other quantum numbers, m and l, describing the states due to movement of the exciton centre of mass in the presence of an external potential barrier (1s, 1p, 1d, ..., 2s, 2p, 2d, ...).

The lowest energy state ($n = 1$, $m = 1$, $l = 0$) is given by

$$E_{1S1s} = E_g - \text{Ry}^* + \pi^2 \hbar^2 / 2Ma^2$$
$$= E_g - \text{Ry}^*[1 - (\mu/M)(\pi a_B^*/a)^2] \tag{3.31}$$

The first exciton is then characterized by a displacement towards higher energies of

$$\Delta E_{1S1s} = (\mu/M)(\pi a_B^*/a)^2 \text{Ry}^* \tag{3.32}$$

This displacement is small compared to Ry^*, as long as $R \gg a_B$. This is what justifies the term 'weak confinement'.

Taking into account the fact that the absorption of a photon can only create an exciton of angular moment equal to zero, the absorption spectrum consists of lines corresponding to states where $l = 0$. The absorption spectrum can then be deduced from the energy values of the excitons, with $\chi_{m0} = \pi m$:

$$E_{nm} = E_g - \text{Ry}^*/n^2 + (\hbar^2 \pi^2 / 2Ma^2)m^2 \tag{3.33}$$

The 'free' electron and hole have the energy spectrum

$$E_{ml}^e = E_g + \hbar^2 \chi_{ml}^2 / 2m_e^* a^2 \tag{3.34a}$$
$$E_{ml}^h = \hbar^2 \chi_{ml}^2 / 2m_h^* a^2 \tag{3.34b}$$

Consequently, the additional energy corresponding to the lowest states of the electron and the hole is

$$\Delta E_{1s1s} = E_{1s}^e + E_{1s}^h - E_g = \pi^2 \hbar^2 / 2\mu a^2 = \text{Ry}^*(\pi a_B^*/a)^2 \tag{3.35}$$

which is much smaller than Ry^*.

3.2.2 Strong confinement

The condition of having a strong confinement is: $a \ll a_B$. In this case, the confined electron and hole have no related states, and the zero-point kinetic energy of the electron and the hole is much greater than the value of Ry^*. The Coulomb interaction between the electron and the hole can be neglected, and we can consider, at a first approximation, that movements of the electron and the hole are not correlated. Therefore, each particle has the energy spectrum given by (3.34). This spectrum is illustrated in Figure 3.6.

The selection rules for optical transitions show that, for the electron and the hole, only transitions between states of the same principal and orbital quantum numbers are allowed. Therefore, the optical absorption spectrum consists of a set of discrete lines given by

$$E_{nl} = E_g + \hbar^2 \chi_{nl}^2 / 2\mu a^2 \tag{3.36}$$

This is the reason why quantum dots at the strong-confinement limit are sometimes called *artificial atoms* or *hyperatoms*.

However, we should not forget that the electron and the hole are confined in a space comparable to that of the exciton of infinite crystal. Therefore, we cannot analyse

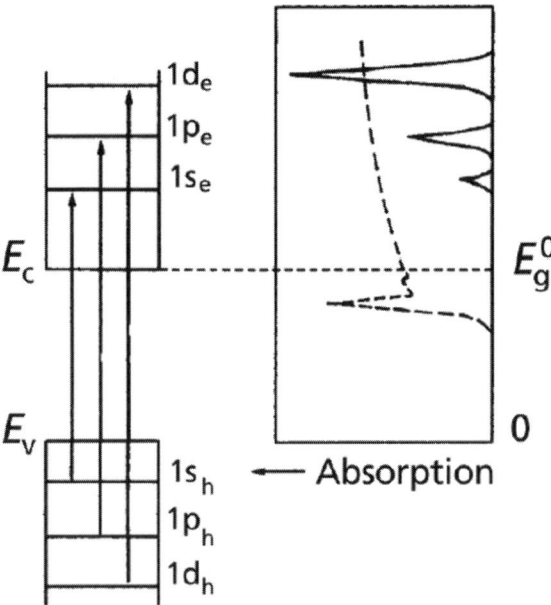

Figure 3.6   *Optical properties of an ideal spherical quantum dot. On the left, the energy levels of electrons and holes correspond to the energies of a particle in a box of infinite inner walls. The optical selection rules allow electrons and holes to have transitions between states of the same quantum number. Therefore, the optical absorption spectrum (on the right-hand side) is only composed of discrete lines while the corresponding absorption spectrum of the bulk material (dashed curve) is continuous*

the electron and the hole independently from each other. In the problem, we must include the two-particle Hamiltonian, including the terms of kinetic energy, Coulomb potential and confinement. It becomes the following expression of the energy of the electron-hole pair's fundamental state:

$$E_{1s1s} = E_g + \pi^2\hbar^2/2\mu a^2 - 1.786e^2/\varepsilon a \tag{3.37}$$

The last term of the right-hand member describes the effective Coulombian interaction. This term is not negligible and is more important than the corresponding term of the bulk crystalline material. This is the main difference between the quantum dot and the one- or two-dimensional system, for which the Coulomb energy of an electron-hole pair is equal to zero.

Taking into account the different terms, the fundamental energy of the exciton is expressed by a shift towards higher energies ('blue shift') of the optical absorption threshold of semiconductors when the nanoparticles' size decreases. Some examples are given in Figure 3.7.

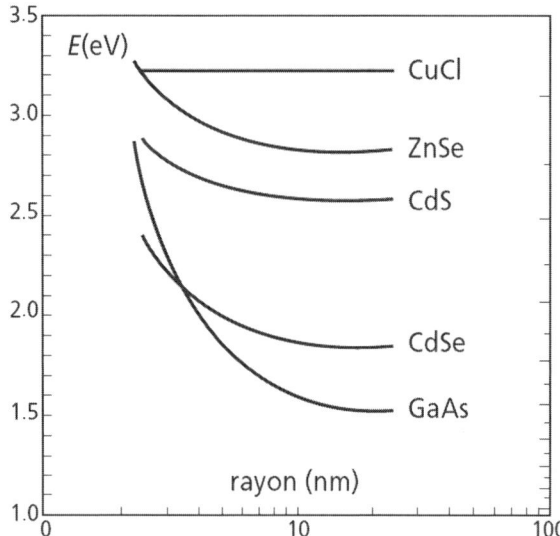

Figure 3.7   Energy of the optical absorption threshold of some semiconductors, regarding the radius of the particle. When the particle's radius is greater than 20 nm, the optical absorption threshold corresponds to the bulk material one. When the radius is inferior to about 2 nm, the effective-mass approximation is no longer always valid

The approach used up until now assumes that the electron and the hole are characterized by properties (effective-mass …) identical to the crystalline solid ones. This approach gives a semi-quantitative idea of properties of nanoparticles. For a more accurate description, it is essential to explicitly take into account the effects of size on electronic properties, as well as the real shape of the particles. In reality, nanoparticles are generally included in a matrix (semiconductor, polymer, etc.), whose effect is to modify the potential barrier at the edge of the particle. The energetic spectrum of the excitons as well as the optical selection rules can be modified.

3.3   Optical properties of metallic nanoparticles

Optical properties of metallic nanoparticles are dominated by the collective excitations of the conduction electrons that are the result of the interaction with the electromagnetic radiation. These properties are detected in particles Au, Ag and Cu, because of the presence of conduction electrons. The electric field of incident radiation causes the appearance of an electric dipole in the particle. To offset this effect, a force is created in the nanoparticle, at a single resonance frequency.

The oscillation frequency depends on many factors, among which are the size and the shape of the nanoparticle, as well as the type of surrounding medium. In the case of non-spherical nanoparticles, the oscillation frequency depends on the electric field orientation. Therefore, two different types of oscillation, transversal and longitudinal, are possible. Moreover, when particles are close enough to each other, interactions between them take place, which modify the obtained results.

Optical properties of diluted systems of spherical nanoparticles, of radius R, are described by Mie theory. The effective cross-section extinction σ_{ext} is given by the relation

$$\sigma_{ext} = (24\pi^2 R^3 \varepsilon_m^{3/2}/\lambda)\{\varepsilon''/[(\varepsilon' + 2\varepsilon_m)^2 + \varepsilon''^2]\} \tag{3.38}$$

where $\varepsilon = \varepsilon' + i\varepsilon''$ is the complex dielectric constant of the metal, and ε_m is the dielectric constant of the surrounding medium.

Diluted systems of nanoparticles are characterized by big variations of colours that depend on the size, the shape and the surrounding medium of the nanoparticles. The explanation comes from the denominator of the previous equation, which predicts the existence of an absorption peak when $\varepsilon' = -2\varepsilon_m$. In this case, the absorption band of the metal is condensed in a narrow band of surface plasmon.

In the case of elongated nanoparticles, we must take into account the light polarization in relation to the symmetry axis of the system.

3.4   Electrical properties: the Coulomb blockade

Nanosystem sizes affect not only energy levels, but also their electrical properties. A nanoparticle crossed by an electron becomes electrically charged. Consequently, the nanoparticle will behave like a capacitor, and its capacitance must be taken into account. These charge effects are characterized by what we call the *Coulomb block-ade*. When we want to add an additional electron in a nanoparticle, we must provide an additional energy to the nanoparticle:

$$E_{add} = \Delta E + e^2/C \tag{3.39}$$

where ΔE is the energy difference between the system energy levels at N and $N + 1$ electrons, and C is the nanoparticle's capacitance. The nanoparticle's behaviour is illustrated by Figure 3.8.

When the potential difference applied between the extremities of the nanoparticle is such that the Fermi level of the source is lower than the energy level corresponding to the particle with $(N + 1)$ electrons, there is no electric conduction throughout the particle. The conduction is possible when the applied potential difference V is enough in a way that the Fermi level of the source is in coincidence with (or above) the particle energy level with $(N + 1)$ electrons.

3.5   Quantization of electrical conductivity

When the mean free path of the electrons (greater than a few tenths of a nanometre at ambient temperature) is much inferior to the nanoparticle's diameter, the classical theory of electrons diffusion can no longer be applied. The theory applicable to

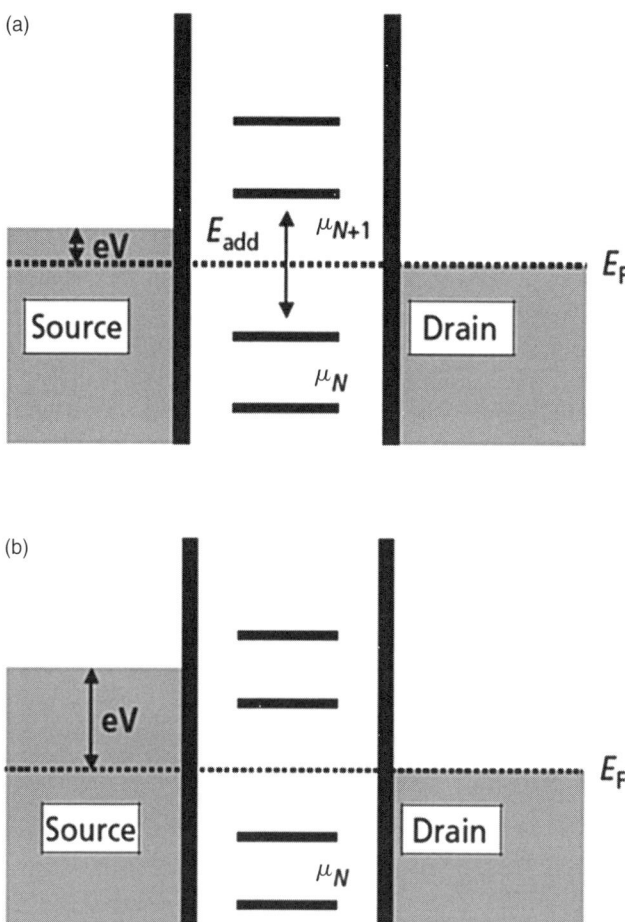

Figure 3.8   Coulomb blockade

nanoparticles was given by Landauer in 1957. The significant result of this theory is that the electrical conductivity is quantized. The quantum of conductivity is given by

$$G_{qu} = (e^2/\pi\hbar) = 8 \times 10^{-5}\Omega^{-1} \tag{3.40}$$

The consequence of this effect is that the I-V characteristic of a nanoparticle (or of a molecular junction) is quantized (see Chapter 4).

The electronic properties of nanosystems are different from those of matter at bigger scales. These properties give nanosystems novel physical and chemical properties.

Chapter 4
Molecular electronics

Gordon Moore, author of Moore's law in 1965, founded Intel, the electronic processor manufacturer, in 1968 with Robert Noyce. In line with his own predictive laws, the first Intel microprocessor (Intel 4004), manufactured in 1971, contained some 2 250 transistors, while the Pentium 4 has more than 42 million transistors on a surface area of only a few square centimetres (Figure 4.1).

A large number of electronic components assembled to create a device that carries out a specific function (e.g. to calculate in the case of a microprocessor or to store in the case of memory) is called an *integrated circuit*. Nowadays, integrated circuits are usually fabricated on silicon wafers by using light beams on each component on the wafer (a process known as *photolithography*). As the years go by, the evolution of computer performance takes place in parallel with a progressive miniaturization of component size, which today has reached a tenth of a micrometre. This progressive increase in integrated-circuit complexity has given birth to the second Moore's law, which states that the manufacturing cost of microprocessors doubles at each generation; the production line of the next generation of Intel processors should cost around €2.5 billion.

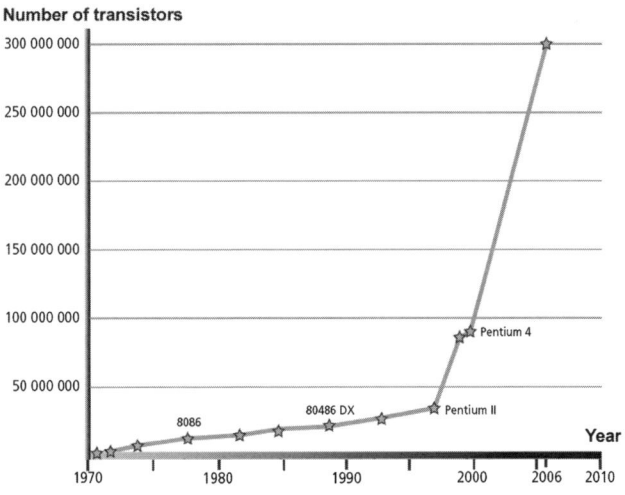

Figure 4.1   Moore's law

The future of silicon may be called into question in the coming years, as it is more and more difficult to develop according to the trend given by Moore's law. This deviation from Moore's law will be ruled not only by economic criteria but by technological limitations as well. Because the resolution of the photolithography process is highly dependent on the wavelength of the beam used, an increased miniaturization of a current circuit implies a reliable source and an extremely complex manipulation of ultraviolet radiation or low-energy X-rays. The current performance forecast for 2010 also suggests that the transistor's size reduction implies that only a few electrons would be responsible for the current switching in the transistor. This lets us foresee the possibility of statistical errors in binary processing (0 or 1) of data. A new approach must be developed to evolve from microelectronics to nanoelectronics and to manufacture tomorrow's computer. One approach is to use molecules instead of silicon; this defines the nascent field of *molecular electronics*. In this context, a lot of research groups all over the world are currently trying to demonstrate that traditional electronic components (such as conducting wires, diodes, transistors) can be made by involving only one molecule or, in a more realistic way, a limited number of molecules.

A molecular approach is very attractive:

- unlike nonorganic semiconductors, organic molecules have properties that are easily adjustable by molecular engineering, by changing the molecule size for instance, or the nature of their framework, or by introducing some substitutes;
- furthermore, molecules have the ability to auto-assemble, that is to say to spontaneously build highly organized structures in their allotted space;
- using molecules can also lead to a significant size reduction of current components of around two orders of magnitude (i.e. a factor of 100).

In this chapter we will describe the emergence of the field of molecular electronics, particularly the different types of components developed at a molecular scale, which can be involved in the integrated circuit of the future.

4.1   Molecular wires

4.1.1   Mechanical junctions

The simplest electronic components of an integrated circuit are the connections that carry the current from one element to another (between two transistors, for instance). Professor Mark Reed's group (Yale University) was one of the first to demonstrate that linear conjugated organic molecules can carry current between two metal contacts and then behave like molecular wires. To this end, a molecular junction formed by two gold tips separated by a variable distance of only a few tenths of a nanometre is realized by controlled mechanical breaking (Figure 4.2). The idea illustrated by Figure 4.2(b) is to locally break a gold filament of a few nanometres in diameter, glued on a flexible substrate, by progressively bending the substrate using a piezo-electric material (material that changes dimensions when a potential difference is

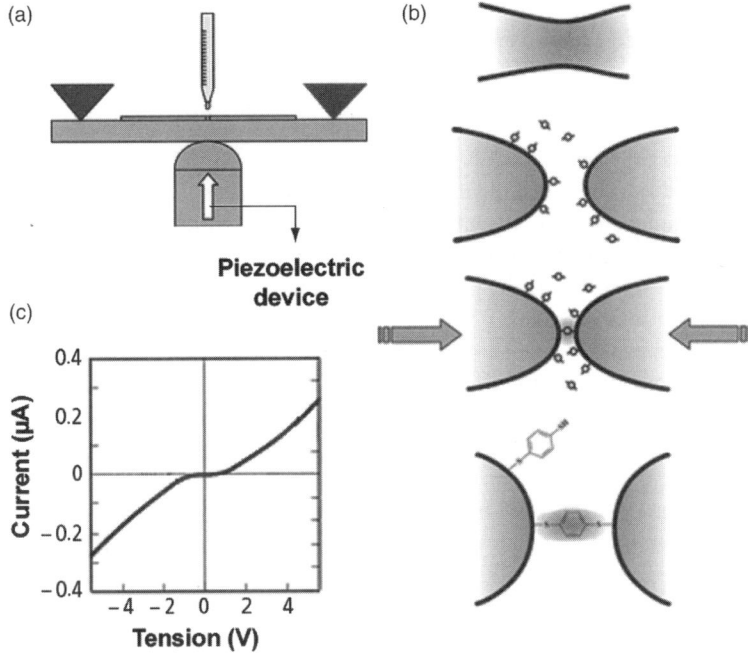

Figure 4.2   *Mechanical junction: (a) development of a mechanical junction using a piezoelectric material; (b) development of a mechanical junction by inserting a benzenedithiol molecule between the two gold tips; (c) the I/V curve generally found for this junction. [Source: based on data published in Reed et al., paper, Science, Vol. 278, p. 252, 1997]*

applied). Benzenedithiol molecules in solution are then brought into contact with the mechanical junction and get adsorbed on the metal tip thanks to the development of a covalent bond between gold atoms and the sulphur atom of the molecule (associated with the elimination of the hydrogen atom of the *thiol* group SH). Finally, in the last step, the size of the molecular junction progressively decreases by a piezoelectric inverse effect until only one single molecule is simultaneously attached to the two metal electrodes (Figure 4.2(b)).

At this stage, it becomes possible to measure the current going through the molecule as a function of the potential difference applied between the two contacts, as illustrated in Figure 4.2(c). This last figure shows that currents of the order of a milli-amp go through the molecule for some potential differences of a few volts; it shows as well that this current does not increase linearly with the applied voltage. Then, electrical properties of molecular junctions deviate from Ohm's law commonly used in traditional electrical circuits, which states that the connection between the voltage V and the current I is a constant known as the resistance R of the circuit:

$$R = V/I \tag{4.1}$$

Experimental results show that the junction resistance is a function of the applied potential difference, and it is in the order of a few megohms (MΩ).

4.1.2   The contribution of high-resolution microscopy

Another smart approach, developed by Professor Paul Weiss's group (Pennsylvania State University), is based on the use of *scanning tunnelling microscopy* (STM). This technique was developed in the early 1980s by Gerd Binnig and Heinrich Rohrer at IBM Zürich, earning them the 1986 Nobel Prize in Physics. The technique consisted in bringing a metal tip (typically platinum and iridium) composed of only a few atoms at its end close to a conducting surface and locally measuring the current between the tip and the surface. The lack of contact between the surface and the tip forced the electrons to cross the gap in order to establish the current between the conductors. This important effect, known as the *tunnel effect*, is heavily dependent on the distance between the tip and the surface; and the tunnel current decreases exponentially as a function of the distance. A piezoelectric reading device measures slight variations with a precision of 0.1 Å (one-hundred-thousandth of a micrometre) – smaller than an atom! The system is linked to a computer, which reconstructs the image of the surface. The STM technique allows the characterization of a metallic surface by analysing, during the scanning of the surface, the displacement of the tip required to measure a constant current.

Within the context of molecular electronics, the STM was used to analyse a surface, above which have been attached some conjugated molecules (an oligomer of paraphenylene ethynylene) surrounded by molecules with a saturated carbonaceous backbone (alkanethiol molecules), illustrated in Figure 4.3. In this case, the quantity of current going through the tip to the surface depends highly on the type of molecules placed between them. If a constant current must be measured between the two conductors, the tip will have to move closer to (or away from) the surface as the

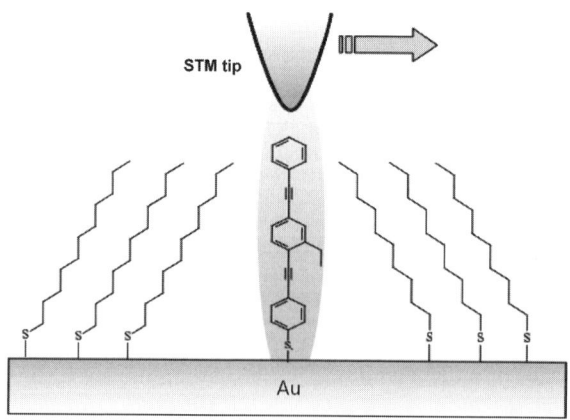

Figure 4.3   Characterization of electrical properties of conjugated molecules by STM

molecule conducts less (or more) of the electric current. This effect is clearly seen by STM, showing that a conjugated system is a better molecular wire compared with a saturated ring.

4.1.3 Current through a molecule

Let us now describe the origin of the current measured in a molecular junction. When a molecule is put into contact with two metal electrodes, the Fermi level E_f of the electrodes (i.e. the energy level at the boundary between the valence band VB and the conduction band CB) is generally situated in the forbidden band FB of the molecule, the band that separates its occupied and empty molecular orbital (Figure 4.4(a)). The occupation probability of the valence band and the conduction band by electrons at a given temperature is ruled by the Fermi function:

$$f(E) = \frac{1}{\exp\left[\frac{(E-E_f)}{kT} + 1\right]} \tag{4.2}$$

We can see in Figure 4.4(b), at 0 K, that the Fermi function is equal to 1 if $E < E_F$ and equal to 0 if $E > E_F$; this implies that the valence band is completely full and the conduction band is empty. However, some electrons can be transferred from the top of the valence band to the bottom of the conduction band at temperatures higher than 0 K. In the case of the molecule, the big energy gap (in the order of a few eV) between the occupied and the empty orbital implies that no occupied orbital is filled by thermal effects in normal temperature conditions.

In order for electrons to go through the molecule, it is necessary that an occupied level of an electrode be in resonance (i.e. having the same energy) with an unoccupied orbital of the molecule, the unoccupied orbital being itself in resonance with an unoccupied level of the other electrode (Path 1 of Figure 4.5). In the same way, the

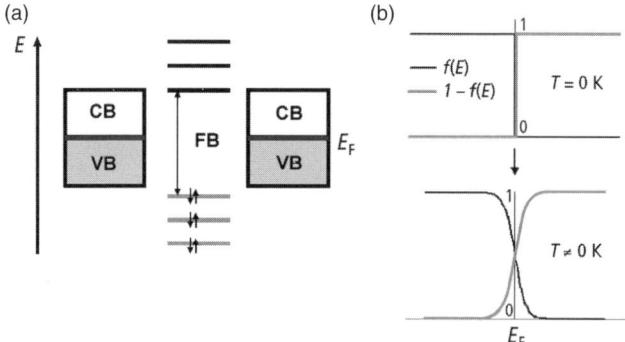

Figure 4.4   *Current through a molecule: (a) electronic structure of a non-polarized molecular junction, with VB and CB the valence bands and conduction bands of the metal, E_F the Fermi level, and FB the forbidden band of the molecule; (b) Fermi function from 0 K to a different temperature*

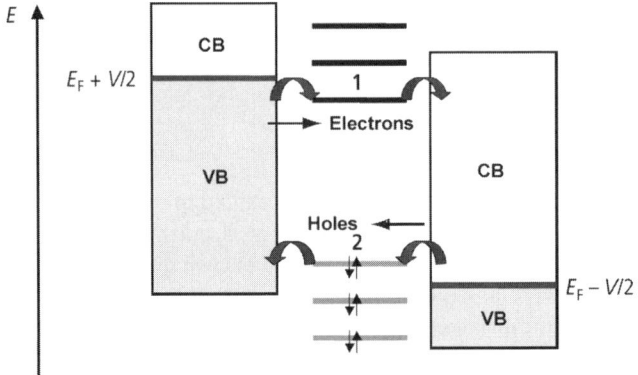

*Figure 4.5   Current transport through a molecule due to the potential difference V
between the two electrodes*

current can also be obtained when an unoccupied level of an electrode is in resonance
at the same time with an occupied orbital of the molecule and an occupied level of
the other electrode, in other words, when a hole (i.e. an unoccupied electronic place)
is initially injected in an occupied orbital of the molecule and then transferred to the
valence band of the second electrode (Path 2 of Figure 4.5). In the light of Figure 4.4,
these two situations are not detected in the case of a non-polarized molecular junction;
this explains the absence of current at 0 V in the experimental measurements. The
only way to generate a current through the molecule is to apply a potential difference
V between the two electrodes in order to destabilize the Fermi level of one electrode
by $-V/2$ (on the positive pole's side of the potential difference) and to stabilize the
Fermi level of the other electrode by $+V/2$ (on the negative pole), as illustrated in
Figure 4.5. Note that this symmetrical evolution of Fermi levels is detected only in
the presence of perfectly symmetrical junctions.

Figure 4.5 illustrates that only the molecular orbitals included between $E_F - V/2$
and $E_F + V/2$ can take part in the transport of electrons (for unoccupied orbitals)
and holes (for occupied orbitals), through the junction. The current going through
a given molecular orbital coupled with the orbitals i and j of the two electrodes is
expressed as

$$I = \frac{4\pi e}{\hbar} f(E_i)(1 - f(E_i))n_i n_j T_{ij}^2 \delta(E_i - E_j) \tag{4.3}$$

where e is the electron charge; $f(E_i)$ is the occupation probability of level i in the
first electrode; $(1 - f(E_j))$ is the inoccupation probability of level j in the second
electrode; n_i and n_j are the number of electronic levels having respectively the energy
E_i and E_j; $\delta(E_i - E_f)$ is the Kronecker symbol (equal to 0 if E_i and E_j are different and
equal to 1 if they are identical), whose presence guarantees the energy conservation
during a transport described as elastic; T_{ij} is a term of transmission that expresses the
coupling force between the molecule orbitals and the two electrodes.

This coupling term depends highly on the bonding force between the sulphur atoms of the molecule and the gold atoms of the electrodes, and on the shape of the molecular orbital of the considered level as well; the transport of the current will be drastically reduced if the orbital is located on a segment of the molecule but will, however, be amplified for a molecular orbital rearranged all along the conjugated path of the molecule. On this basis, the current can be obtained by summing the contribution from each orbital of the molecule; it will be identical whatever the polarity in the case of a perfectly symmetrical junction, symmetry regarding the shape of the molecule as well as the type of contacts at the metal/molecule interface. All asymmetry will lead to a different current for two voltages of same value but of opposite polarity, therefore bringing a rectification process, which will be described in the next section.

This very simple view of current transport that has just been described is at the root of Landauer's theory commonly used nowadays to calculate the current through a molecular junction as a function of the voltage. This approach takes into account all the complexity of the problem, due to the fact that the coupling elements as well as the energy and the shapes of the molecule orbitals generally vary as a function of the applied potential difference; note that the orbital's energy is less affected by the electric field only if the molecule is perfectly symmetrical and has identical contacts with the two electrodes.

4.1.4   Coulomb blockade

When a very small contact is made between the molecule and the electrodes in the absence of covalent bonds, the current/voltage (I/V) curves generally have the form of stairs, which is the expression of a phenomenon known as *Coulomb blockade* (Figure 4.6). In this case, the current is not a direct current (DC) between the two electrodes and each charge injected into the molecule remains there a certain time. In the case of a transport dominated by electrons, the first step of the graph corresponds to the injection of a charge into the molecule. The insertion of a second charge cannot be made so easily because of a phenomenon of electrostatic repulsion between the two new electrons injected into the molecule. This is the reason why the current remains

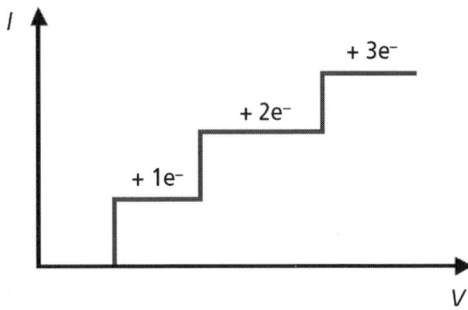

Figure 4.6   *I/V curve representing the transport of electrons throughout a molecular junction in presence of a Coulomb blockade*

constant until the potential difference offsets the repulsion energy between the two charges. At this stage, two electrons can be simultaneously injected, which makes the amplitude of the current increase at a value that will remain constant until three electrons can be simultaneously injected. Generally, the effects of Coulomb blockade are more noticeable in the I/V curves, as the temperature at which the experiments are conducted is low.

4.2   Molecular rectifiers

In 1974, Doctor Ari Aviram (IBM Research, Yorktown Heights) and Professor Mark Ratner (Northwestern University) proposed for the first time the theoretical concept of a molecular rectifier, equivalent to the diodes used currently in electrical circuits (to transform alternative current into direct current, for instance). This discovery has earned them the title of the founding fathers of molecular electronics. The mechanisms of these rectifier devices are expressed by the asymmetry of the I/V curve; this asymmetry illustrates that the current measured for a given potential difference greatly varies with its sign, as illustrated in Figure 4.7.

The principle of obtaining molecular rectifiers suggested by Aviram and Ratner consists in placing, in a metal junction, a molecule composed of a conjugated segment of a π-type electron donor, connected to a segment of a π-type electron acceptor, via a short saturated ring. In this molecule, the characteristic of the acceptor segment is expressed by a high stabilization of its HOMO and LUMO energy levels (*highest occupied molecular orbital* and *lowest unoccupied molecular orbital*) compared with the energy levels of the donor segment. Therefore, it is conceivable that the HOMO levels of the donor and LUMO of the acceptor are both close to the Fermi level of the metal electrodes (Figure 4.8).

Assuming this situation of equilibrium, the application of a small potential differ-ence in the *forward bias condition* (direct polarization, $V > 0$) allows the modulation

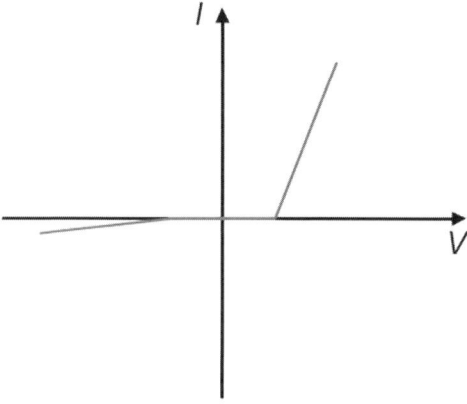

Figure 4.7   Simplified representation of the phenomenon of rectification in a diode

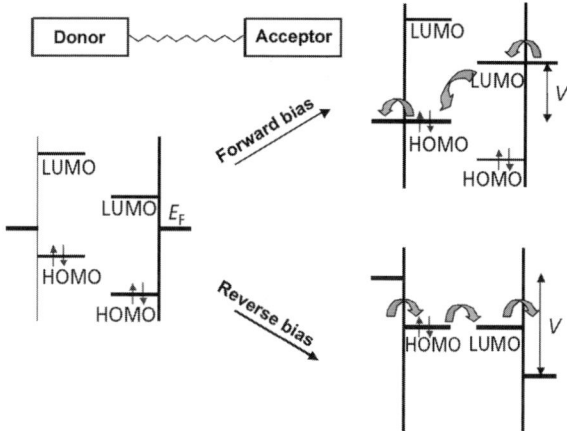

Figure 4.8   Molecular rectifier: electronic structure of a junction including a molecule made of a segment of π-type electron donor and a segment of π-type electron acceptor, with the transition of the current through the molecule for a forward bias polarization and a reverse bias polarization

of the Fermi levels of the two electrodes in order to inject an electron into the acceptor's LUMO and to remove an electron from the donor's HOMO (i.e. to inject a hole). This mechanism generates a current through the junction by transferring the injected electron of the acceptor's LUMO into the vacant space left in the donor's HOMO. We can observe that the energies of the HOMO and LUMO levels of the two segments progress as a function of the applied potential difference, in the same direction as the Fermi level of the nearby electrode. This behaviour is explained by the fact that each segment is located close to an electrode and therefore does not have a perfect symmetrical position in the molecular junction.

On the other hand, a potential difference far superior to the one needed in direct polarization is required in reverse polarization to generate a current. This current is due to the activation of a resonance between the donor's HOMO and the acceptor's LUMO. This asymmetry is then the cause of the rectification expected for that kind of device (Figure 4.8).

It was only in the early 1990s that the groups of Professor Roy Sambles (Exeter University) and Professor Robert Metzger (University of Alabama) discovered a molecule that could perform the rectification suggested by Aviram and Ratner: the hexadecylquinolinium tricyanoquinodimethanide molecule (Figure 4.9). A rectification ratio close to 26:1 (corresponding to the intensity ratio of the current for the two polarizations) has been obtained for this molecule at 1.5 V at ambient temperature. However, in the light of recent research we have to admit that the Aviram–Ratner mechanism is not the cause of the obtained rectification; actually, the rectification is due to the asymmetrical positioning of the molecule's conjugated part within the junction because of the presence of the alkyl ring $C_{16}H_{33}$. Moreover, the measured current for a monolayer was really small ($<10^{-6}$ A). This expresses the difficulty of

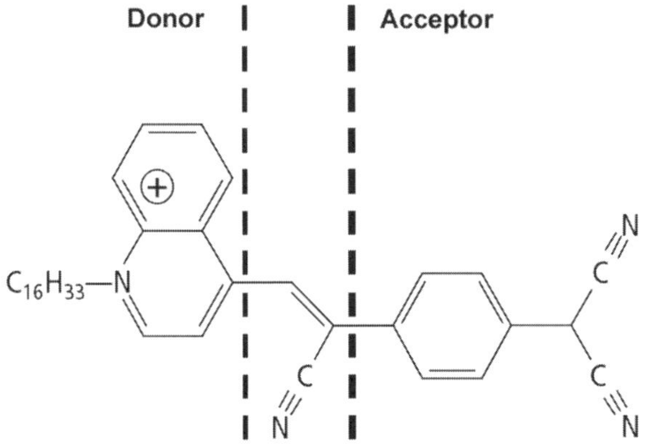

Figure 4.9   Chemical structure of the molecule hexadecylquinolinium tricyanoquinodimethanide

transporting the current through the conjugated part of the molecule, demonstrating therefore the small interest aroused by this molecule in the nanoelectronics field. Most of the rectifications analysed in current devices are due to the absence of symmetrical contacts on the electrodes or due to the asymmetry of the molecule within the junction. A lot of research is being conducted on new molecular structures leading to high rectification ratios and high current.

4.3   Molecular transistors

Transistors are currently the primary elements of integrated circuits, where they are combined to create logic gates that are used to process the received digital data. Unlike diodes, transistors have three electrodes. In the 'field effect' configuration, transistors are usually made by depositing an insulating layer over a first metal electrode (known as *gate*), followed by the deposition of a semiconducting layer contacted by two other metal electrodes (known as *source* and *drain*) (Figure 4.10). A transistor works like a switch by controlling the electrical current going through the source and the drain due to the potential difference applied between the gate and the source. Performances of transistors are judged according to the ratio of current measured in their on and off states (generally higher than 10^6), as well as the switching speed between one state and the other; this switching depends on the mobility of charge carriers in the semiconductor. Integrated-circuit speed performances of data processing are governed by this switching speed. In the molecular electronics field, the most remarkable recent developments are the ones of Professor Cees Dekker's group (Delft University, Netherlands), which was the first to achieve the fabrication of a molecular transistor based on a carbon nanotube (Figure 4.10).

Carbon nanotubes were developed for the first time in 1991 by Dr Sumio Iijima (NEC Fundamental Research Laboratory). These tubes are made of a great number of merged benzene cycles; they look like a graphite layer that has been rolled up. The

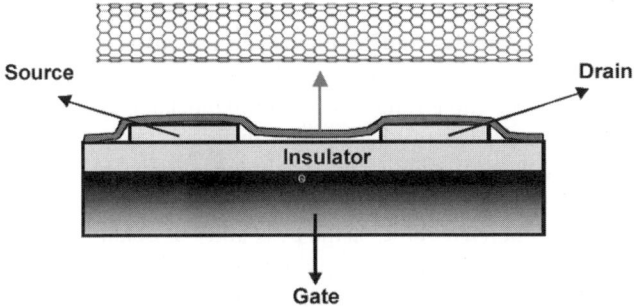

Figure 4.10   Chemical structure of a carbon nanotube and schematic architecture of a transistor

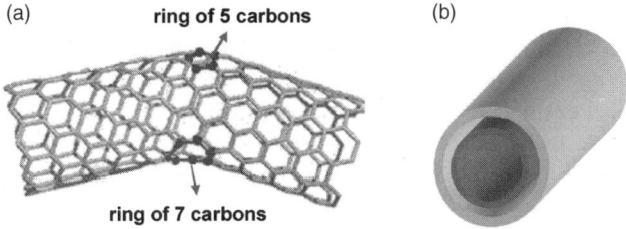

Figure 4.11   Formation of an intramolecular junction within a nanotube (a); structure of a multi-walled nanotube (b)

diameter of this kind of tube generally varies from a nanometre to a few micrometres. Several tubes can also fit into one another like Russian dolls, therefore leading to multi-walled nanotubes (Figure 4.11). These materials currently arouse a lot of interest because of their mechanical properties (they have a Young modulus greater than the steel one, expressing therefore their high mechanical resistance) and their electrical properties. Consequently, carbon nanotubes can be metallic or semiconductor, depending on their diameter and the way the graphitic sheet is rolled up.

Building the first molecular transistor based on a carbon nanotube working at ambient temperature and using classical lithography techniques, Professor Dekker's group first created two platinum electrodes (*source* and *drain*) over an insulating layer of SiO$_2$, deposited over the device's *gate* made of doped silicon. The carbon nanotube was finally deposited between the two electrodes by means of a solution. In this device, the quantity of current passing between the source and the drain through the nanotube (on a distance of 140 nm corresponding to the distance between the two electrodes) could be highly modulated by applying a potential difference between the source and the gate. The demonstration of this transistor effect (the nanotube can be switched between an insulating and a conductor state) was a major breakthrough in the molecular electronics field, therefore allowing the ambition of further miniaturization of current integrated circuits.

The interest aroused by carbon nanotubes is not limited to the transistor effect. These materials are also good candidates for the fabrication of molecular wires. Recent experimental studies show that molecular orbitals are situated along the tube (this favoured charge transportations) and that Coulomb blockade exists at a low temperature (with a difference of a few mV between the different steps of current). Moreover, because of their high thermal stability, the maximum current a nanotube can transport without dissipation is far greater than the one achieved by copper or even gold wires. Another benefit of carbon nanotubes is the possibility of obtaining a rectifier device by joining a metal and a semiconductor within the nanotube; that kind of junction can be obtained by intentionally introducing chemical defects (a ring of five or seven carbons) within the nanotube (Figure 4.11).

However nanotubes also have some drawbacks. Their synthesis usually leads to a large mixture of tubes of different diameters and electrical properties with different numbers of tubes fitted into one another. It is therefore essential to characterize and sort out all the tubes before using them in a system; this is a serious limitation in the case of development into high-volume production. However, this problem is avoided for conjugated molecules of small size, which can be synthesized in several billion identical exemplars. This is the reason why many current research works aim to involve those molecules within transistors of nanometric size. We should also note that the appropriate position of the nanotube over the electrodes is not always guaranteed during the fabrication of the transistor; it requires, for instance, the use of tips of high-resolution microscopes in order to move the tubes to the required position. Then, it will be important in the future to develop techniques allowing semiconductor materials (molecules or nanotubes) to be easily inserted in the space between the electrodes.

4.4   Molecular resonant tunnelling diodes

The development of traditional electronic components (diodes or transistors) at the molecular scale has demonstrated the possibility of creating unachievable devices in traditional microelectronics, like resonant tunnel diodes. These diodes are molecular junctions characterized by the presence of a high and narrow peak of current in the I/V curve (Figure 4.12). This behaviour is different from the common curve's profile, which is an increase in the current for an increase in the potential difference. This characteristic, known as *negative differential resistance* (NDR), is likely to be used within a new type of electrical switch where the transition between the forward and reverse biased condition is induced by the potential difference applied at the junction terminals. This unusual behaviour has been detected for the first time in inorganic material-based devices that also use resonance effects between energy levels and charge transfers by tunnel effects (see below); hence the terminology of *resonant tunnel diodes*.

The first molecular resonant tunnel diode was demonstrated by Professor Mark Reed's group by fabricating a nanopore (cavity drilled within an inorganic material by ion bombardment) connected at its extremes by two gold electrodes; the

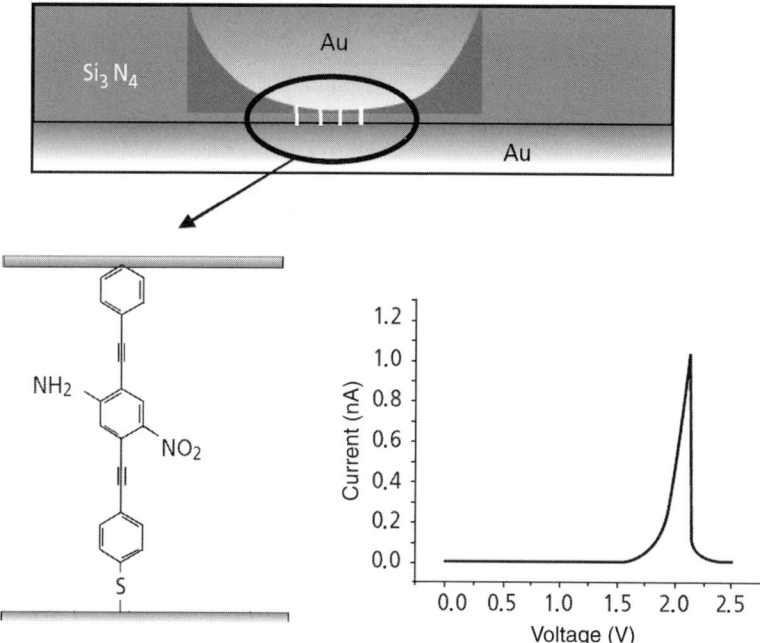

Figure 4.12   *Formation of a nanopore and NDR behaviour. Schematic representa-*
tion of the formation of a nanopore and NDR behaviour observed for
phenylene-ethynylene oligomer. [Source: based on experimental data
published by Reed et al. (Science, Vol. 286, p. 1550, 1999)]

three-ring phenylene-ethynylene oligomers were substituted by nitro and amino
groups (Figure 4.12). At 60 K, this device shows a high NDR peak at 2 V, associated with a current a thousand times higher than the current measured at around the
peak (1 nA *vs* 1 pA). The NDR behaviour does not last at ambient temperature for
this molecule. However, this behaviour is observed at ambient temperature for the
molecule having only a nitro group in the central ring. Conversely, no peak of current
is observed, whatever the temperature, for the molecule that is not replaced in the
central ring and for its derivative, having only one amino group.

The origin of this peak of current is not yet well understood; many different interpretations are given within the scientific community. One plausible interpretation is
that molecular orbitals are not delocalized all along the conjugated backbone. Consider, for instance, a molecule where respectively the LUMO level and the LUMO+1
level are located on the left-hand side and right-hand side of the molecule, as depicted
in Figure 4.13; and let us assume that the potential difference allows the electron to
go through the molecule from left to right. The potential difference will allow us to
inject an electron in the molecule's LUMO, without generating a high current within
the junction, because there is no charge relocation. The increase in the voltage leads
to a destabilization of the LUMO level and a stabilization of the LUMO+1 level,

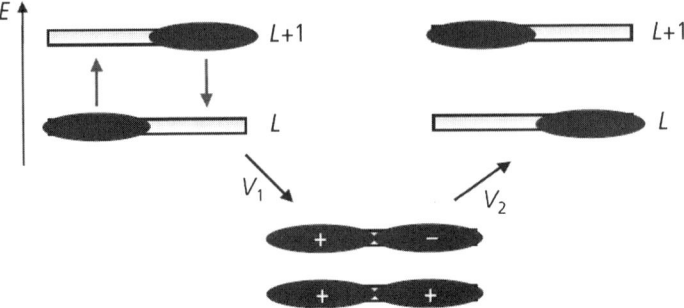

Figure 4.13   Possible alignment mechanism of molecular orbitals due to a potential difference, leading to NDR behaviour

until reaching a critical stage where the energy of the two levels become equal. At this stage, the two levels combine with each other to form two orbitals completely delocalized on the molecule. Therefore this situation favours an effective carrying of the current through the junction. This resonance between the two levels will vanish with a further increase in the voltage, therefore leading to a drop in the current.

A high current will then be observed for a narrow range of potential, therefore decreasing the probability of finding a peak in the I/V curve. In the case of phenylene-ethynylene oligomers, a localization of the orbitals is caused by the rotation of the central ring (which can be due to thermal effects or the interaction between the molecular dipole moment and the electrical field generated by the potential difference). Then, the NDR behaviour can be the result of orbital alignments around the external rings of the molecule, with the current going through the centre ring by a tunnel effect. Only more detailed analysis will allow us to validate this method, which combines resonant effects with tunnel effects, like equivalent devices based on inorganic material.

4.5   Molecular memories

A memory circuit is an electronic component based on the arrangement of different elements (typically transistors and capacitors), allowing it to store data via electrical charges, to read and to erase on demand. Thanks to the recent work of Professors James Heath and Fraser Stoddart (University of California, Los Angeles), we know that those functions can be implemented at a molecular scale. This has been discovered by inserting a mono-layer of rotaxane molecules between two metal electrodes, and measuring the I/V curve of the device. A rotaxane molecule is typically composed of a long linear chain along which a molecular ring can slide.

As an example, the rotaxane molecule presented in Figure 4.14 is composed of a linear chain including an element A resulting from the derivative of tetrathiafulvalene molecule, an element B of dioxynaphtalene, and a molecular ring positively charged (four times) and composed of bipyridine elements. In the neutral state, the molecule

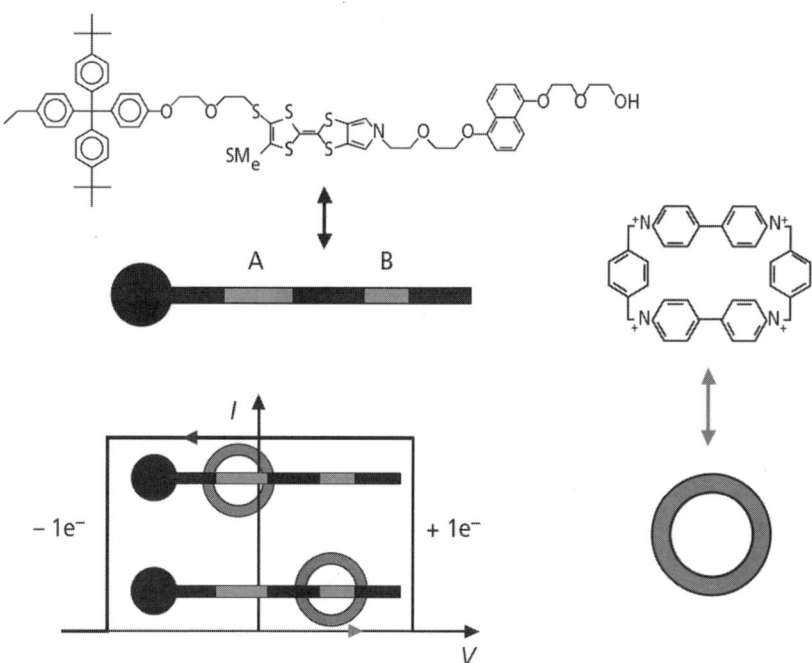

Figure 4.14   *Rotaxane molecule and memory principle. Chemical structure of a rotaxane molecule, and illustration of the mechanism of memory for a monolayer between two metal electrodes. Those results have been presented in depth by Stoddart* et al. *in* Accounts of Chemical Research, *Vol. 34, p. 433, 2001*

is characterized by a high interaction between the ring and the element A and by an important current through the molecular junction. A negative voltage oxidizes the molecule by removing an electron from element A, which results in the displacement of the molecular ring towards element B, due to Coulombian repulsive forces between the ring and element A, which is now positively charged. This displacement is associated with an important drop of the current, therefore turning the device, evolved from a direct biased condition, into a reverse bias. The current is restored only after a positive voltage is applied, leading to the reduction of element A and the inverse displacement of the molecular ring (Figure 4.14).

We can then see that it is possible to write data that change the electrical conductivity of the system and erase them by restoring the original current. Moreover, because the forward and the reverse biased conditions are both likely to be observed over a large range of intermediate potentials (known as the *hysteresis phenomenon*, which expresses the fact that a scanning of the voltage from right to left or in the opposite way does not give the same I/V), the state of the system (i.e. the eventual written data) can be read at any time. This molecular system then achieves the three principal functions of an electronic memory.

4.6   Towards the molecular computer

All the elements that compose a current integrated circuit can now be made on a molecular scale. Does it mean that tomorrow's computer will be molecular? This cannot be taken for granted. Only future developments will bring us the answer. Certainly, many more problems must be resolved before the development of such a computer can be realized. It is then necessary to develop techniques, for instance to connect (efficiently and in a reproducible way) the individual elements, to avoid interference between components, to minimize the heat dissipation and to connect molecular circuits to macroscopic devices. Other competing technologies are likely to supplant this nascent molecular electronics field: for instance, some devices where the data medium is not so much the electron charge as its spin (spintronics field) or a photon (photonics field) or even a quantum state (quantum computing field). The 'nano' processing of data is then a really complex field, and predicting its future is therefore very difficult.

Chapter 5
Neuroelectronics

5.1   When electronics meets biology

Within the integrated circuits of computers (which we can compare to 'artificial brains'), the data transmission between the different elementary elements – the transistors – is made by electrical impulse. It is interesting to note that within the human brain, and between the brain and the other different organs, transmission of information (a function fulfilled by the nervous system) is also based on electrical phenomena. The sending of information all along the system of nerve cells (neurons) is expressed by the propagation of a change in the electrical charge of the neurons' membrane. In this way, an electrical impulse goes through the nervous system to the organs. The electrical characteristics of nerve impulses were demonstrated two centuries ago, during a well-known experiment in which Volta placed two electrodes on frogs' legs and applied a voltage, in this way producing a movement. But it is only recently that the similarity between the transmission process of information in microelectronics systems and the nervous systems made researchers think of correlating transistors and neurons. This new scientific field, which relates microelectronics concepts and neurology, has been named *neuroelectronics*.

The first point of interest in this approach (making transistors and neurons communicate with each other) is that an integrated circuit associated with a neuron (or a group of neurons) forms a sensor of the neurons' activity. Then, those systems would be able to monitor *in vitro* (within test tubes) the effect of new medicine on the nervous system, which will allow the reduction of *in vivo* (in-the-body) tests (on animals).

The functioning of the human brain (a network of more than a hundred billion neurons) is far from being completely understood. To connect a neuron network to a microelectronic integrated circuit would allow the network to be observed while it was working and then to better understand the functioning of our brain; moreover, it would allow the use of this network to assist the computer in some tasks and calculations. This is the second major aspect of neuroelectronics.

Finally, we can also imagine that an artificial microelectronics system could restore the communication in an area of the nervous system damaged after an illness or an accident. Those artificial neurons could be used for instance to repair spinal-cord injuries after some accidents. Of course, all those applications exist currently only in the speculative world of researchers, but the recent advances of neuroelectronics (a scientific field only ten years old) are promising. In the next sections these progresses will be briefly presented.

5.1.1   Communication between neurons and transistors

The main aspect in neuroelectronics is the communication between the primary element of integrated circuits, the transistor, and the primary element of the nervous system, the neuron. In other words, research in this field has at least two aims. On the one hand, the sending of an electric signal by the integrated circuit must generate an action potential in the neuron (i.e. the electrical impulse that characterizes the passing of nervous information), and, on the other hand, the passing of an action potential through the neuron leading to a change in the electrical properties of the transistor. Professor Fromherz and his group (Max Planck Institute, Martinsried, Germany) were the first to establish this communication between transistors and neurons. In 1991, German researchers showed that the emission of an action potential by a neuron in contact with a field-effect transistor (the principle of this transistor is described in Chapter 6) could induce a change in the gate potential, therefore a modification of the current in the transistor channel. The measurement of this current allows the neuron's activity to be monitored. Then in 1995, the same group generated an action potential in a neuron from an electrical impulse delivered by a capacitor controlled by a transistor within the integrated circuit. They demonstrated that neurons and transistors can communicate.

5.1.2   Neurons' control over integrated circuits

Having surmounted this first stage, in order to build more complex systems, it is necessary that the localization and the growing of neurons on top of integrated circuits be well controlled. If the development of neurons is not controlled, the dendrites (thin branched projections of a neuron that act to conduct the electrical stimulation from other neural cells to the cell body) will grow randomly. Moreover, neurons have a tendency to move at the surface during their development. It is, however, mandatory that neurons and dendrites be placed in particular places of the integrated circuit (for instance, on the electrodes used for the stimulation). To this end, two approaches are possible: the biochemical and the topographical approach.

5.1.2.1   Biochemical method

The biochemical method uses a lithography process to place at the appropriate regions of the surface a thin layer of protein, with which the external part of the neuron's membrane has a great affinity. Therefore, the neurons will grow preferentially in the region covered by the protein (Figure 5.1).

5.1.2.2   Topographical method

In the topographical method, a polymer layer is deposited all over the surface of the integrated circuit. Then, regions of a few micrometres of diameter are hollowed out

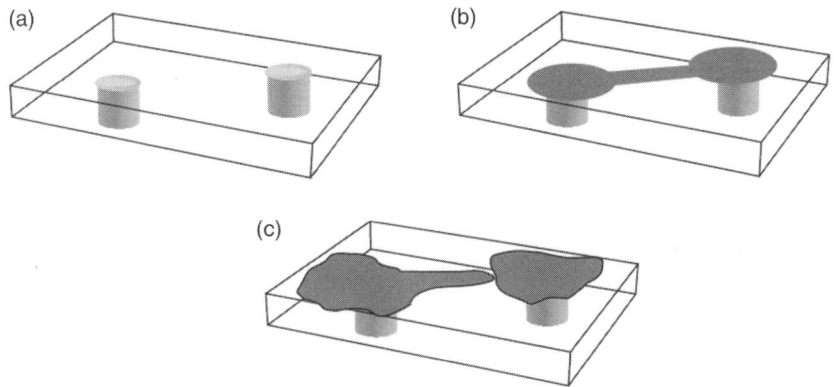

Figure 5.1   *Positioning control of neurons by a layer of spatially structured proteins: (a) the integrated circuit is composed of two transistors (the two blocks); (b) between the blocks is deposited a thin layer of proteins, represented by the grey zone; (c) the neurons are localized and they develop in a region defined by the protein layer*

at appropriate places by a lithography process. Those 'wells', linked by trenches also made using the lithography process, will then receive the nerve cells. Dendrites will then develop in the trenches, establishing the communication between the neurons located in different wells (Figure 5.2).

A variant of this second approach consists in depositing, on the surface, polymer blocks of 40 μm height and a few μm in diameter. The polymers are placed in a circle; they make an enclosure within which the neuron is confined. The dendrites grow within those blocks and they will join the neurons placed in the next enclosures.

By these different methods, it is then possible to generate networks of functional neurons (neurons communicate between themselves) in a precise geometry imposed by the structure of the integrated circuit.

5.1.3   Electronic circuit between two neurons

Finally, the most recent progress in neuroelectronics is the connection of two neurons separated by a microelectronic circuit. This was also made by Professor Fromherz and his group. The system is depicted in Figure 5.3. A first neuron is placed on a field-effect transistor (FET), whose function is to detect the neuron's activity. The electronic circuit measures the electrical conductance variations in the transistor, measuring the action potential of the neuron. In response to an action potential of the first neuron, the circuit generates an impulse stream that is applied to the second neuron via stimulating electrodes. Those impulses then cause the generation of an action potential in the second neuron (this action potential is detected by implanted microelectrodes in the neuron). The analysis of this hybrid system shows that the activity of the second neuron is directly correlated to the activity of the first neuron;

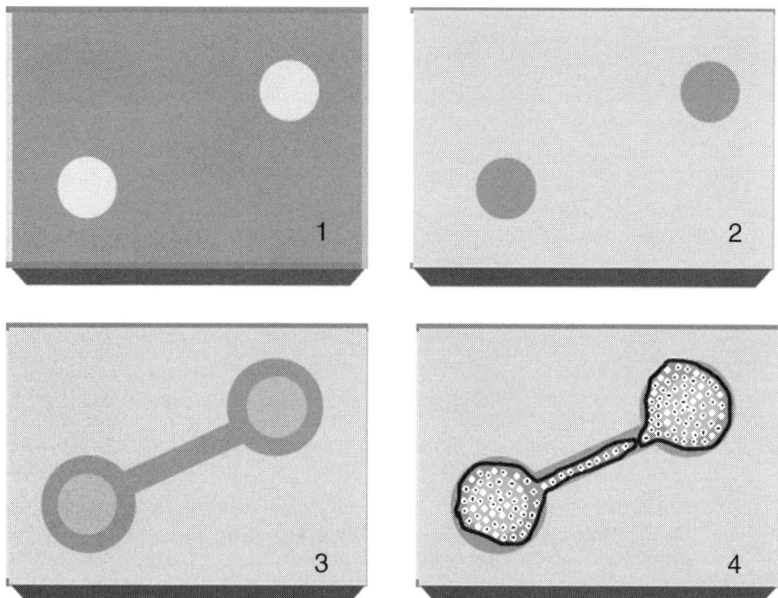

Figure 5.2   Positioning control of neurons by topographical confinement. The surface of the depicted integrated circuit contains two transistors, represented by the circles (1). A polymer layer is then deposited (2). This layer is etched by lithography in order to expose the transistors and to create trenches between the wells (3). The neurons are confined in the wells and dendrites grow in the trenches (4)

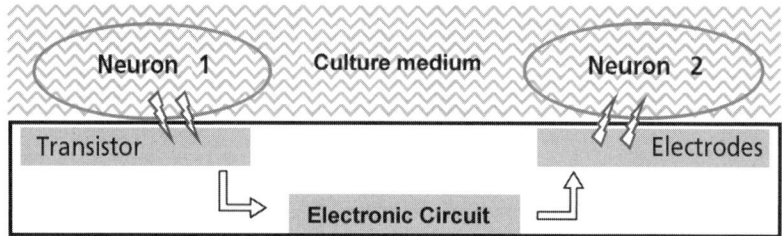

Figure 5.3   Communication system of neuron – integrated circuit – neuron. The emission of an action potential by the first neuron (represented by the downward flashes) is detected by the transistor; the electronic circuit then generates an impulse stream (represented by the upward flashes) applied to the second neuron via stimulating electrodes

in other words, the two neurons communicate via the electronic circuit. Then, the first stage of the development of neuroprothesis (artificial neuron) is surmounted.

This progress opens up the way to the integration of neuronal networks in microelectronic systems for the development of hybrid computers within which living material and inert nanostructured material will closely work together. Also fascinating is the prospect of the use of DNA (the medium of biological information) in information technology.

5.2   A computer based on the DNA double helix

Beyond the miniaturized silicon chip and the molecular electronics, strange and futuristic concepts of tomorrow's computers exist. The DNA computer is one example. The macromolecule of DNA, composed of two polynucleotide chains entwined around each other as a double helix (demonstrated by two biologists, James D. Watson and Francis H. Crick, in 1953), is the storage and transmission system of the genetic information within cells. Using the initial letters of the four nucleic acids (adenine, thymine, cytosine, guanine), the DNA encodes, at a molecular level, all hereditary information.

DNA chains pair in a double helix according to strict rules of complementarity (an adenine pairs only with a thymine via hydrogen bonds, an adenosine with a guanine; see Figure 5.4a). Moreover, many billions of chains can be held in a single tablespoon. By applying a logical operation to each strand (for instance, by exploiting

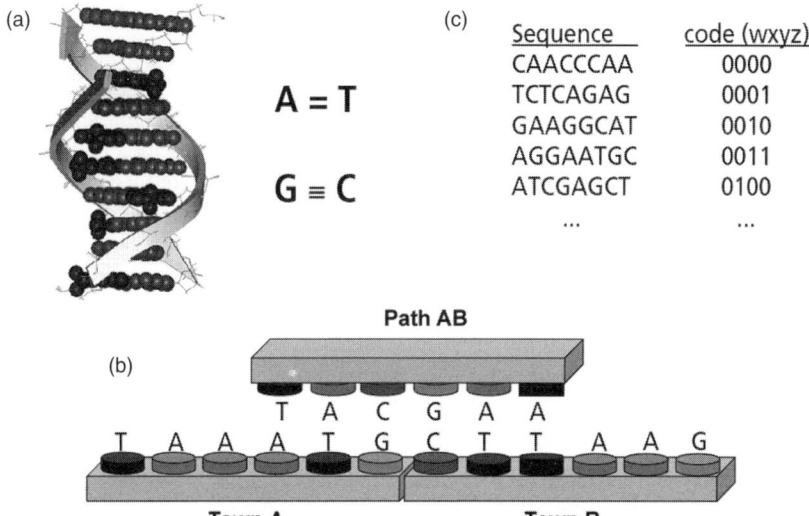

Figure 5.4   *The DNA computer: (a) complementarity rules between bases (A = adenine, T = thymine, G = guanine, C = cytosine) and three-dimensional structure of a double helix DNA; (b) schematic representation of the coding of towns and paths by complementary DNA strands in the 'problem of the sales representative'; (c) the other way to solve the problem of the sales representative, by coding the information with specific sequences*

the strands' complementarity – see below), scientists can foresee the construction of a 'super-computer' capable of processing billions of operations simultaneously. Therefore, the DNA computer can be seen as a super-computer where the huge numbers of parallel DNA strands play the role of 'nanoprocessors'.

In 1994, the first important step towards the development of the first DNA computer was made by Professor Adleman from the University of South California. He used DNA to solve the 'problem of the sales representative' – finding the shortest route between a certain number of towns, but visiting each town only once. In the case where the number of towns is small, the problem can be easily solved manually. However, when the number of towns increases, the number of potential routes increases exponentially, making the resolution of the problem very difficult, even with a computer.

Adleman solved the problem by using a few grams of DNA in such a way that all the solutions were generated simultaneously. The different steps used in his algorithm are shown below.

- Random paths are generated between all towns. Towns are coded as simple DNA strands of specific sequence; the paths are coded by strands whose first half is complementary to a town while the second half is complementary to another town (Figure 5.4b). All those oligonucleotides are then mixed in a solution, and combine with each other according to the rule of complementarity, thus generating a great number of random solutions.
- The paths that satisfy the problem are selected. This step, which is the longest one (seven working days in the laboratory), consists of identifying the hybridization corresponding to paths that join the departure and arrival towns by passing through all the other towns only once. A huge assortment of complex molecular-biology tools (extraction by means of gels, use of magnetic blocks, etc.) has been used to this end.

Other researchers suggested that a DNA computer, based on strong structures where DNA strands are fixed on a metallic surface rather than floating in the solution, could be more efficient. Therefore, DNA fixed on a gold surface has been used to solve a problem similar to that of the sales representative. In this problem, the solution must satisfy simultaneously a certain number of logic conditions, each of them composed of many variables. The procedure is as follows. Each element of information is coded as an oligonucleotide of a particular sequence (see Figure 5.4c). First, all DNA strands that correspond to all the possible solutions are fixed on a gold surface. Then, oligonucleotides that can satisfy the first condition are added and pair with certain strands at the surface, in this way creating some DNA double-strands at the metal surface. The remaining DNA single strands, which do not satisfy the condition, are destroyed by enzymes. The surface is then heated to clean out the complementary DNA segments, then the surface is cleaned and new DNA strands that satisfy the second condition are deposited and pair with the complementary strands at the surface. This cycle is repeated for each condition until, at the end of the process, only the sequences that satisfy all the conditions remain on the surface.

Huge technical barriers still need to be surmounted before the first DNA computer can be developed. One of the major problems faced is that DNA is a complex structure that comes with errors, which can lead to wrong results from the computer. In nature, errors are corrected automatically by our cells, while currently there are no correction methods for the DNA computer. Another difficulty, brought up earlier, is the need to do a huge number of complex operations to 'read' the final solution (even if the generation of all the possible solutions is 'instantaneous'). In practice, it is therefore unlikely that in the near future the DNA computer will be used to do common tasks that can be fulfilled by conventional computers. Suggestions for the use of DNA computers include the creation of biosensors that could identify pathogenic agents in a medium or detect biochemical processes within the cell as well as research within databases for genetic applications.

Chapter 6
Plastic electronics

Most known polymer materials behave like good electrical insulators; therefore, most sheaths of electric cable are made of polypropylene. However, a specific group of polymers known as *conjugated polymers* reveal amazing electrical properties. Thanks to a simple oxido-reduction reaction known as *chemical doping*, electrical properties of conjugated polymers can be increased by a few orders of magnitude and can even become comparable to the conductivity of some common metals such as iron and copper, which is suggested by the term they are known by: *synthetic metals*. As a matter of interest, *Synthetic Metals* is the name of an international scientific journal that publishes works done on conducting polymers, and Professor Alan Heeger (co-winner of the 2000 Nobel Prize in Chemistry) has long been its principal editor. The interest for conducting plastics is that, in a single material, the electrical properties of metal can be combined with the mechanical properties of polymers (plasticity, lightness, ease of manufacture, low cost, etc.) This potential leads to significant advances in the development of conducting polymers for technological applications. Some of them have been achieved, e.g., antistatic plastic films that cover photography films produced by Agfa-Gevaert SA.

6.1   Conjugation in conducting polymers

Before describing the electronic structure at the root of the properties of conjugated polymers, let us first trace the brief history of polymers. Even if since 1960, many research works suggested that polymers could have significant electrical conductivity, the discovery in 1973 of the metallic nature of polymeric sulphur nitride, $(SN)_x$ (an inorganic polymer), was the first step in the development of conducting polymer materials. The electrical conductivity of $(SN)_x$ is around 10^3 S/cm, compared with 10^5 S/cm for copper and 10^{-14} S/cm for polyethylene. It is worth noting that polymeric sulphur nitride becomes superconductive (its electrical resistivity drops to zero) under critical temperature of 0.3 K. In the mid-1970s, it was demonstrated that $(SN)_x$ electrical conductivity can be increased by an order of magnitude by exposing the polymer to bromine vapour or other oxidizing agents. In this case, the polymer material is in a polycation form, the charge neutralization being achieved by adding the reduced form of the oxidant (Br_3^- in the case of exposure to bromine).

It is the adaptation of this chemical oxido-reduction to the polyacetylene (organic conjugated polymer, the structure of which is described in Figure 6.1) that leads to the first decisive progress in the field of conducting plastics. In 1977, Professors

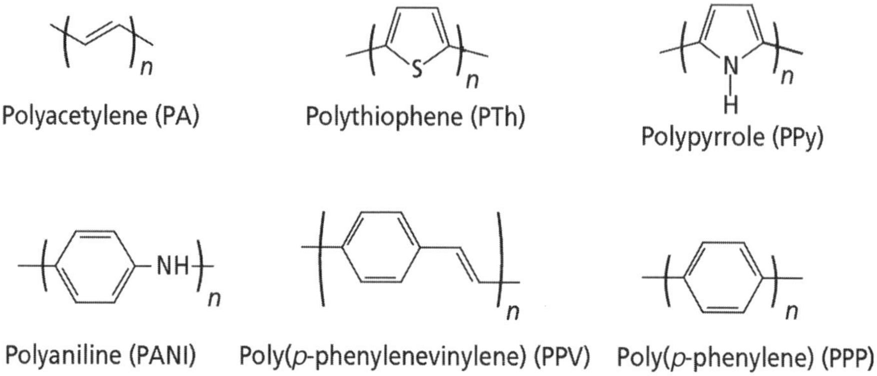

Polyacetylene (PA)   Polythiophene (PTh)   Polypyrrole (PPy)

Polyaniline (PANI)   Poly(*p*-phenylenevinylene) (PPV)   Poly(*p*-phenylene) (PPP)

Figure 6.1   Chemical structure of some conjugated polymers

MacDiarmid, Heeger and Shirakawa discovered that polyacetylene, which has intrinsic conductivity lower than 10^{-5} S/cm, increases its conductivity by eight orders of magnitude (the conductivity reaches 10^3 S/cm) after the exposure of the material to an oxidizing or reducing agent. It is important to underline that this discovery was made thanks to the existence, since the end of the 1960s, of thin films of polyacetylene with good mechanical properties (while, previously, polyacetylene was in powder form, insoluble and infusible). This discovery earned its discoverers the 2000 Nobel Prize in Chemistry.

Afterwards, the same oxido-reduction was applied to many other polymer materials in order to improve their electrical properties and especially their photochemical stability (polyacetylene deteriorates rapidly by oxidation in ambient air). The chemical structures of principal conducting polymers are presented in Figure 6.1. In the 1980s, much progress was realized in the solubilization of conducting polymers in common organic solvents by the substitution of principal chains by alkyl groups or other groups. This progress was decisive in the development of the materials: they were usually made from a solution of polymers. It is for instance the case for 'spin-coating', whereby a solution is deposed on a spinning plate; the spreading of the solution and its simultaneous evaporation lead to the development of homogeneous thin films.

Another remarkable discovery was announced by the group of Professor Friend in 1990: the first light-emitting electrical device (a *light-emitting diode*, or LED) based on a conjugated polymer. Once again, the combination of properties inherent to plastic materials and inherent properties of conjugated polymers (in this case, their optical characteristic, precisely their luminescence) made the materials very attractive for many applications. In parallel to their use in display systems, conjugated polymers can also be used as active components in devices as diverse as FETs and photovoltaic cells. In all those devices, described in depth later, the operating mode is based on the intrinsic semiconducting nature of the conjugated polymer. In order to understand the electrical, optical, and magnetic properties of conjugated polymers, it is essential to analyse carefully their electronic properties; which is the subject of the next section.

6.2 Electronic structure and electron–phonon coupling

A conjugated polymer is a macromolecule in which the backbone is composed of carbon atoms or heteroatoms, each having a π-type atomic orbital, which is the case for all the structures presented in Figure 6.1. The overlap of those π-type atomic orbitals leads to the development of π-type molecular orbitals delocalized all along the chains; this gives conjugated polymers their special properties. In the ideal case of infinite conjugate chains, those molecular orbitals are close regarding their energies and form continuous electronic bands, as predicted by the model of a 'particle in a box' (see Chapter 2); they are known as *type-p electronic band structures*. The highest occupied electronic band is known as the *valence band*; the lowest unoccupied one is the *conduction band*. These two bands are separated by a forbidden band: the forbidden zone. In conjugated polymers in their neutral state, the width of the forbidden band is around 2 to 3 electronvolts (eV), which puts them in the category of semiconductors or insulators.

From this simplified description, one can understand that conjugated polymers can easily be oxidized or reduced; this doping or oxido-reduction reaction (chemical or electrochemical) allows the polymer (an insulator or a semiconductor in its neutral state) to be converted into a doped or charged state, where the polymer is a conductor. Type-p electrons can easily be removed or added to conjugated polymer chains, the latter having low ionization potential. Moreover, the doping reactions do not affect the cohesion of the chemical structure, essentially ensured by the σ-type electrons.

Another aspect essential to the understanding of electrical conduction mechanisms is the high coupling that exists between geometrical and electronic structures, known as *electron–phonon coupling*. The existence of such a coupling is at the root of the terminology used to describe charged species or neutral excitations in conjugated systems. Consequently, the insertion of an electron (or a hole) in a conjugated chain – which means the reduction (or the oxidation) of the chain – induces a local deformation of the lattice, leading to a geometrical defect known as *polaron* in solid-state physics. In the language of chemists, a polaron corresponds to a radical-ion (cation if the charge is positive, anion if it is negative) associated with a local geometrical deformation (of size around 30 Å) of the chain. A bipolaron (two charges of same sign confined in the same region of the chain) is the result of the insertion of a second charge. It comes with stronger geometrical modifications.

These geometrical modifications lead to a rearrangement of the electronic structure; consequently electronic levels appear in the initial forbidden band. These electronic levels are involved in new optical transitions, which are expressed by deep changes in the visible region of the absorption spectrum, i.e. changes in the material's colour. In fact, this colour change, due to electrochemical doping, is at the root of the use of some conjugated polymers in electrochemical devices. The local chain relaxation and its effect on the electronic structure and the optical properties are illustrated

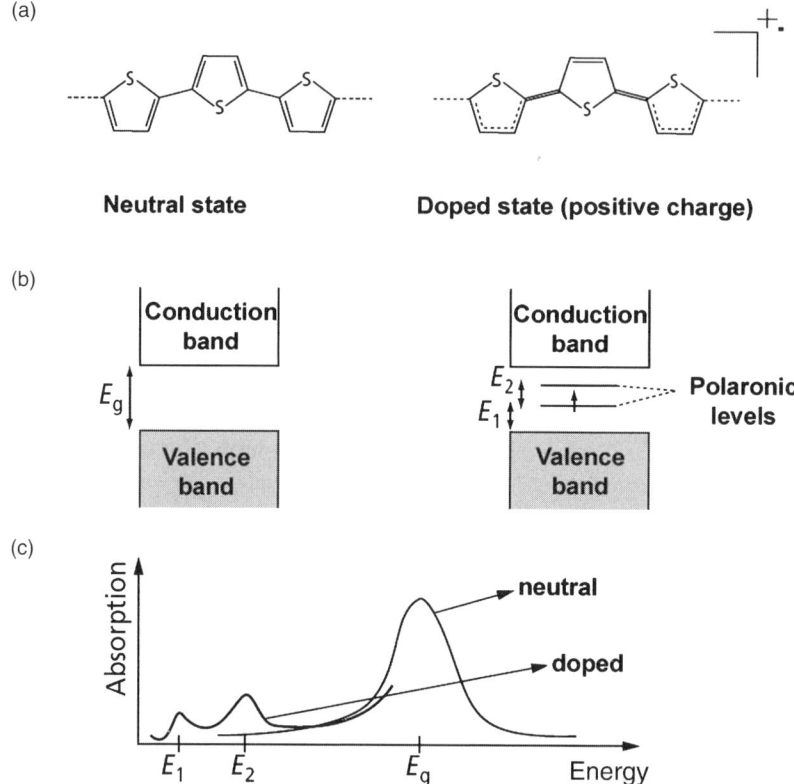

(a)

Neutral state   **Doped state (positive charge)**

(b)

(c)

Figure 6.2   *(a) Geometrical deformations due to doping in a polythiophene chain;*
(b) electronic structure; (c) absorption spectrum of the polymer at
neutral and doped state

in Figure 6.2, for the case of a chain of polythiophene doped by oxidizing agents. It is
worth noting that it is the electronic structure's rearrangement, induced by the insertion of charges, that allows the thermodynamic stabilization of the bipolarons, whose
development is penalized by the Coulombian interactions (two charges of same sign
repulse each other). This can be explained by the fact that the geometric and electronic
relaxation associated with the formation of bipolarons leads to a stabilization of the
system, by the appearance of electronic levels within the forbidden band, which can
offset the electrostatic repulsion.

Finally, it is important to underline the differences between the mechanisms of
charge generation described above for conjugated polymers and inorganic doped
semiconductors. In those materials, atoms are bonded in the three dimensions by
covalent bonds. The lattice is therefore less deformable than a polymer chain (nearly
one-dimensional) and the models of rigid bands generally give a good description of
their properties in the solid state. The doping process is obtained by the insertion of a

tiny quantity (in the order of a few ppm) of atoms such as boron or phosphor, which lie in the lattice by substitution or in an interstitial way. In the example of boron-doped silicon, boron has one valence electron less than silicon (three electrons instead of four). This generates empty electronic levels associated with boron slightly above the silicon valence band. These empty doping levels can be filled by thermal agitation of electrons coming from the valence band, generating holes on top of the band and activating the possibility of electric conduction. By comparison, in a polymer chain, electronic levels within the forbidden band are the intrinsic levels of the polymer material, while the doping agent's intrinsic levels are generally located outside the polymer's forbidden band and they are not actively involved in the conduction mechanisms.

6.3   Charge transport

The interest in conjugated polymers was initially created because of applications that use their properties as good electrical conductors. However, polymers are of current interest due to the fact that, when used in structures where short oligomers or molecules are involved, they have semiconductor properties (they are used at a neutral state in devices). An important mechanism regarding the efficiency of the device is the electrical charge transport within the organic lattice.

The nature of transport mechanisms within materials composed of small conjugated organic molecules at the neutral state remains the subject of many research works. Two different models are generally used depending on experimental conditions. A first transport mode is based on a band model. In this case, the HOMO levels (*highest occupied molecular orbital*) of each molecule interact to form a valence band of intermolecular type (unlike the valence band described in the previous section for a conjugated polymer), while the LUMO (*lowest unoccupied molecular orbital*) levels interact to form a conduction band (Figure 6.3). In this mode, holes and electrons are shared in a great number of molecules (in other words, their wave functions are delocalized on a larger number of molecules).

The main parameter characterizing the transport is the mobility of the charge carriers (μ), defined as the ratio between the charge velocity (v) and the amplitude

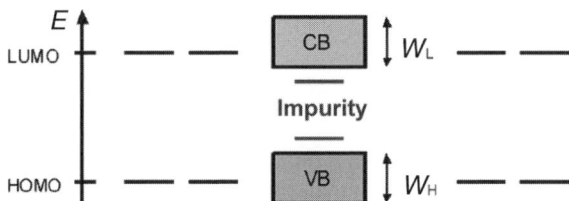

Figure 6.3   Illustration of the development of an intermolecular valence and conduction band for the transport in a band-like regime. The appearance of electronic levels is due to the presence of an impurity within the forbidden band

of the applied electrical field (F), which induce the displacement: $\mu = v/F$. The mobility can be measured in different ways. A widely used approach is the *time-of-flight* (TOF) technique, for which the organic material is sandwiched between two metal electrodes. Charges are generated in the organic lattice by irradiation close to an electrode; then the holes or the electrons migrate through the organic material according to the direction of the potential difference applied between the two electrodes; then the transit time (t) is measured. With knowledge of the amplitude of the electric field and the thickness of the film (d), the mobility can be estimated as $\mu = d/tF$. The obtained mobility is often affected by the presence of defects or impurities; it will then correspond to a lower limit, which will be the dominant one in the system. Another approach is to extract the mobility from the electrical characteristics of a transistor based on the material of concern (*infra-vacuum*); these measures also lead to lower limits. The upper limit can be reached by a more complex technique, known as *PR-TRMC* (*pulse radiolysis time-resolved microwave conductivity*). In this method, the sample is bombarded by highly energetic electrons to create charge carriers, and the variation of microwave absorption is measured and can be directly related to the carrier's mobility.

The mobility is high in a 'band'-like regime (up to 10^2 cm^2/Vs) and is directly related to the width of the valence band (W_H) for the holes and to the width of the conduction band (W_L) for the electrons; these widths can reach a few hundred meV. The larger the width (which expresses a high interaction between molecules), the higher the mobility. In a band-like regime, mobility increases when the temperature decreases, as in metals, because vibrations of the organic lattice decrease at low temperatures, consequently limiting the collisions between the charge carrier and the molecules' nucleus. At high temperature, these collisions scatter the charges in all directions and slow down their movement through the material. At very low temperatures, a dramatic drop of mobility is often observed because of the presence of impurities. The latter is often unavoidable and sometimes induces some electronic levels within the material's forbidden band, as illustrated in Figure 6.3.

If an electron falls to such a level, it will not have enough energy at low temperatures to get out of the level, which explains the electrical-current drop (proportional to the density of charge carriers and their mobility).

The band model is not always acceptable because a lot of parameters favour the localization of the charge on a molecule, for instance:

1. **Temperature**: The temperature will increase the vibration of the lattice and its deformation flexibility, leading to a reduction of mobility, i.e. the effective width of the bands. At high temperatures, the energy obtained by delocalizing the charge (the LUMO of a band is lower than the LUMO of a single molecule, and vice versa for the HOMO) cannot compensate for the energy got by localizing the charge (see 2 and 3).
2. **The electron–phonon coupling**: When the charge is localized, the system acquires energy by changing the molecule geometry to form a polaron, as described earlier.

*Figure 6.4   Cartoon representing the two extreme cases of charge transfer in con-
jugated systems: hopping model (illustrated on the right by a carrier
hopping between two localized sites) and band model (illustrated on the
left by a charge carrier travelling rapidly through the electronic band)*

3. **The polarization effects**: A charge is stabilized in a material for the creation
   of dipoles induced in the subsequent molecules. These polarization effects are
   stronger when the charge is confined, which favours its localization.
4. **The molecular disorder**: In a disorganized system, the HOMO/LUMO energy
   levels of molecules can become very different, either because different structures
   coexist for the molecule or because the environment is different for each molecule.
   These energy-level variations tend to localize the orbitals on a single molecule.
   This evolution can be compared to those observed when going from H_2 molecule
   (symmetrical HOMO/LUMO levels) to HCl molecule (HOMO level centred on
   the hydrogen, and LUMO on the chlorine). Bringing different molecules together
   tends to localize the orbitals.

When the charges are localized, they migrate through the material by hopping from
molecule to molecule, using the *charge-hopping mechanism* (Figure 6.4). The latter
is generally operational in devices because of the disorder inherent in the production
of most organic materials. In contrast, a band model will be effective only at low
temperature for a crystal system without impurities; the nature of transport in an
organized system at high temperature is still debatable and can be halfway between
the two described models.

In a hopping-like regime, the charge mobility is lower compared with a band
regime, because the charge transfer involves an important rearrangement of the system. This is due to the geometry of molecules. Molecules have different geometries
depending on whether they are charged or neutral, which implies a change of their
geometry during charge transfer. Moreover, when the charge is transferred, the polarization of the surrounding system must be modified consequently. The charge transfer
rate can be expressed by a first approximation using Marcus theory:

$$k_{hop} = \frac{2\pi}{\hbar} t^2 \frac{1}{\sqrt{4\pi \lambda k_B T}} \exp\left[-\frac{(\lambda + \Delta E)^2}{4\pi \lambda k_B T} \right] \tag{6.1}$$

where t is the transfer integral describing the interaction forces between the two molecules; this parameter decreases exponentially with the distance between the two molecules. λ is the rearrangement factor that describes the change in the molecule's geometry (internal part) and the polarization of the surrounding environment (external part) during the transfer. ΔE contains two contributions: (i) the energy obtained during the transfer thanks to the electric field (or the energy to acquire if the electric field opposes the transfer); and (ii) the energy difference between the two molecules' HOMO and LUMO induced by the disorder.

The charge mobility in a hopping model depends highly on the order within the material (if ΔE is high in Equation (6.1), the transfer rate k becomes low, which slows down the mobility). It can vary from 10^{-7} cm^2/Vs to 10^{-1} cm^2/Vs within organic lattices according to the degree of order (10^{-2} being the threshold value required for an application). The mobility increases with the temperature and the electric field; both of them can help to overcome energy barriers due to the disorder. The mobility also increases with the density of charge carriers because the levels related to impurities become filled and then are inactive at the concentration of high charge carriers.

Because of the nearly one-dimensional nature of most conjugated polymers (interactions between monomers are much stronger along chains than between chains), the charge transport is more efficient along the chains (via a mechanism equivalent to a band model) than between them. The electric current flow along isolated molecules can be used to achieve wires and transistors at nanometre scale (see Chapter 4, related to molecular electronics). In devices of characteristic dimensions superior to the polymer chain size (typically around a tenth of a micron), the current flow necessarily implies the migration of charge carriers between subsequent chains via a hopping mechanism (polymers are generally amorphous or semi-crystal and consequently not very organized); this inter-chain process is the step that restricts the transfer of charges. In the presence of some chemical or structural defects that are often present along polymer chains, a hopping mechanism is also required to ensure the intra-molecular transport.

6.4   Electronic excitations and optical properties

In the previous sections, the electronic structure of conjugated polymers in their fundamental neutral and doped state – regarding their charge transport properties – was discussed. In this section, the focus will be on the use of those polymers in applications that use their optical properties, i.e. mainly LEDs, photovoltaic cells and lasers. The implementation of these devices involves the conception of electronic species in a neutral, excited state. In an inorganic semiconductor, the excited electronic state simply corresponds to configurations created by the crossing of an electron from the valence band to the conduction band; this excitation does not involve any significant relaxation of the three-dimensional rigid lattice. The optical absorption spectrum then displays a large band, in which threshold is defined by the energy difference between the top of the valence band (or HOMO level) and the bottom of the conduction band (or LUMO level). The energetic gap between HOMO and

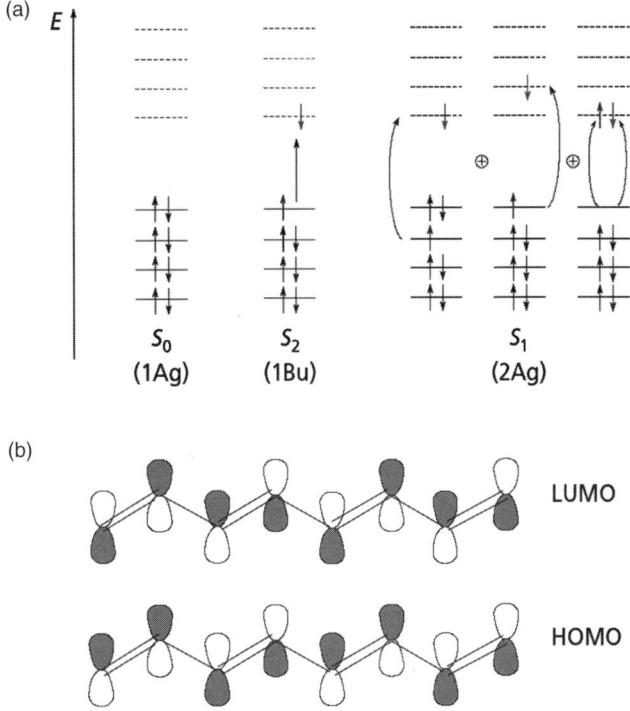

Figure 6.5 *Electronic structure of the octatetraene molecule: (a) electronic con-*
figuration in the fundamental state and the two first excited states; (b)
wave function of HOMO and LUMO orbitals

LUMO, i.e. the forbidden band, sets the colour for which the material absorbs or emits light.

The description is more complex in the case of conjugated polymers, because of the importance of Coulombian interactions and electron–phonon coupling. As an illustration, let us consider the example of the octatetraene molecule (acetylene oligomer or four-unit polyene) described in Figure 6.5. The molecule's electronic structure is characterized by the presence of eight π-type molecular orbitals, whose symmetry alternates between even (*gerade* or symmetric) and odd (*ungerade* or anti-symmetric), and the energy increases with the number of knots in the wave function. The eight π-type electrons are distributed between the eight molecular orbitals, each distribution defining what is called an *electronic configuration*. The electronic configuration of lowest energy (the fundamental state) corresponds to the situation where the π-type electrons occupy two by two the four π-type orbitals of lowest energy.

Because of the strong electron–phonon coupling, excited states and fundamental states generally have distinct geometries at the equilibrium state. In the fundamental state, the molecule geometry is characterized by an alternation of short bonds (double) and long bonds (single), corresponding to, respectively, high and low electronic-bond densities. It is interesting to note that this alternating geometry is expressed in the bonding/anti-bonding nature of the HOMO wave-function: bonding interactions for the double bonds, and anti-bonding for the single bonds (Figure 6.5). Therefore, we can easily understand that an electronic excitation from the HOMO to the LUMO comes with an important rearrangement of the electronic density and consequent geometrical modifications.

The nature of the electronic excitation in conjugated polymers is still being discussed today. Two models are frequently used to explain the experimental results: (1) a band model similar to the one predominant in inorganic semiconductors such as silicon, where the electronic excitations are completely delocalized; and (2) an 'excitonic' model in which the Coulombian interactions combine with a geometry relaxation to induce a localization of the excited states in a restricted spatial domain (the term *exciton* refers to an electron–hole pair bonded by Coulombian interactions and associated with a local geometrical deformation of the lattice). In fact, these two models essentially differ by the value of the exciton-bond energy E_b, which is the difference between the energy of a rigid band-to-band electronic transition and the real energy of the excited state: $E_b \leq kT$ in the extreme case of a band model and $E_b \gg kT$ for an 'excitonic' model. In most electroluminescent conjugated polymers, the exciton-bond energy corresponding to the first excited state (responsible for emitting light) has been estimated around a few tenths of E_v (for instance, $E_b \approx 0.4$ eV in the PPV), which expresses a spatial expanse of around 30–40 Å. This relatively important localization of electronic excitations is due to their low dielectric constants and the low screening of resultant Coulombian interactions (screening refers to the dampening of the electric field by mobile charges).

Another demonstration of the importance of interactions between conjugated polymers is related to the nature of triplet electronic excitations in those materials. If the Coulombian interactions were negligible, singlet and triplet excitons would be degenerate, since the energy gap between those species is proportional to the exchange term. And yet many recent experimental results show an energy gap between the first singlet (S_1) and triplet (T_1) states superior to 0.5 eV in a wide range of conjugated polymers. Besides the fundamental aspects related to the nature of electronic excitations in conjugated polymers, those results have important repercussions regarding the quantum yield of plastic LEDs. In those diodes, the positive and negative polarons' recombination leads to the creation of excitons of the two spin-multiplicities (singlet and triplet). In the case of organic materials, characterized by low spin–orbit coupling, light emission is produced by radiation fallout of singlet excitons, whereas the triplets become de-energized only by a nonradiative process, therefore limiting the electroluminescence quantum yield.

The description of electronic excitation analysed so far is based on a model of isolated polymer chains, a model rigorously applicable in the case of diluted solutions where contacts between molecules are rare. At the solid state (films, crystals), intermolecular interactions can no longer be neglected and profoundly change the nature of the excited states and the resultant optical properties. Figure 6.6 illustrates this situation in the case of two superimposed chains in a cofacial arrangement. As a result of the interaction between the chains, the first singlet excited state (responsible for light

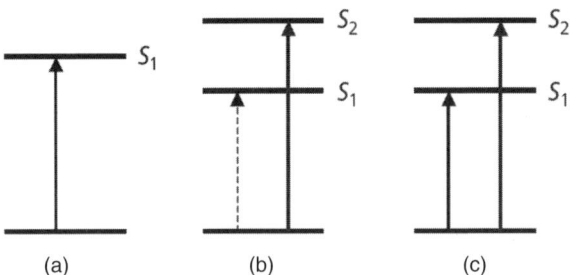

*Figure 6.6   Singlet excited states (Sn): (a) singlet excited state of an isolated chain;
(b) singlet excited state of a dimmer composed of two chains in cofacial
interaction; (c) and forming an angle between their long axes. The
continuous arrow represents permitted electronic transitions while the
dashed one represents forbidden transitions*

emission) of the isolated molecule splits up into two new states characterized by a
wave function delocalized over the two conjugated segments. In the case of a cofacial
configuration, the electronic transition from the fundamental state to the first excited
state is impossible because of the symmetry, and all the intensity is concentrated in the
superior state. Consequently, going from a situation of 'isolated chains' to a situation
of 'chains in interaction' comes with: (1) a displacement towards the highest energy
of the optical absorption spectrum; and (2) a decrease in the luminescence quantum
yield according to the Kasha rule.

As in the case of charged species, the localization/delocalization of electronic
excitations over the subsequent chains is defined by the trade-off between the intermolecular interaction amplitudes (which favour the excited states' delocalization)
and the geometry relaxation over a chain (which tends to localize them). It is worth
mentioning that, unlike the charge transport, the disorder in conjugated materials is
in principle favourable to the luminescence efficiency, since it tends to confine the
excited states.

The situation is, however, more complex because the effective generation of neutral excited species that are emissive in diodes requires a high mobility of charge
carriers (polarons) within the conjugated material. In order to obtain high-yield luminescence in a well-organized medium, the conjugated chains can be substituted – for
instance, using bulky steric groups to reduce the interactions between conjugated
systems. Another possible method, but more difficult to implement, consists in defining molecular structures leading to a stack favourable to the emission: for instance
it is the case when the chains pile up by forming an angle between their long axes
(Figure 6.6).

6.5   Plastic electronics

The initial interest in conjugated polymers was based on applications using their
properties as good electrical conductors; however, this interest has been progressively

diverted to devices involving polymers as well as short oligomers or molecules, and using their semiconducting properties. In the following section, the three more promising applications in this field of *plastic electronics* will be analysed, i.e. light-emitting diodes, photovoltaic and solar cells, and organic transistors in integrated circuits and biochemical sensors.

6.5.1   Organic light-emitting diodes

A light-emitting diode, or LED, is a device that emits light when a current goes through it due to a potential difference applied in direct polarization. The first electroluminescence signal in an organic material was noticed in 1960 for an anthracite crystal, but interest in this phenomenon began in 1987 when Drs Tang and Slyke (Kodak Research) discovered a high green emission from the molecule of tri(8-hydroxyquinoline) aluminium AlQ_3. Three years later, Professor Richard Friend's group (Cambridge University) presented the first electroluminescence signal from a conjugated polymer, paraphenylene polyvinyl (PPV) (Figure 6.7).

The easiest way to create an organic diode from conjugated polymers consists in deposing a transparent metal electrode on a substrate (usually glass). This first electrode (the anode) is typically based on ITO: an oxide composed of indium and tin. The anode can also be composed of a doped conjugated polymer like polyethylene dioxythiophene (PEDOT) mixed with chains of sulphonate polystyrene (PSS) acting as anti-ions. Then a layer of plastic based on conjugated polymers is deposited generally by centrifugation from a solution. Finally, a second metal electrode (the cathode)

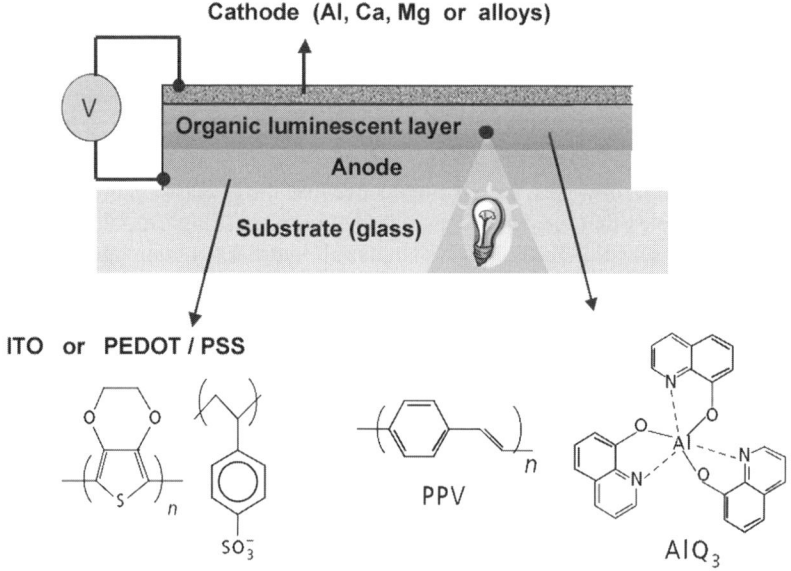

Figure 6.7   Architecture of an organic LED based on conjugated polymers

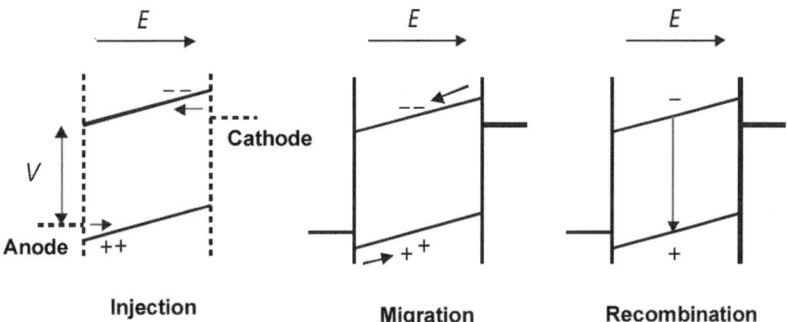

Figure 6.8   *Electroluminescence of an organic diode: illustration of the different stages (injection/migration/recombination) that rule the electroluminescence phenomenon within the directly polarized organic diode*

in calcium, aluminium, magnesium or an alloy of these metals is deposited by evaporation. An architecture composed of many layers is generally used for devices based on molecules.

The two metals sandwiched together with the plastic layer are not arbitrarily chosen: the metal of the cathode must have a Fermi energy level close to the LUMO level of the conjugated chains and the anode one must have a Fermi energy level close to the HOMO. The sequence of four different mechanisms allows the organic diode to emit light due to the potential difference applied in direct polarization. The latter establishes a potential energy gradient of HOMO and LUMO levels through the organic layer, as illustrated in Figure 6.8.

However, very small leakages of current are measured for a similar potential difference applied in reverse polarization, which frequently leads to a rectification rate superior to 10^3.

The first mechanism involved is a charge injection process (electrochemical process of oxido-reduction) which consists in transferring electrons, first from the cathode to the LUMO of the chains, then from the HOMO to the anode (i.e. creation of *holes* within the plastic layer). As the Fermi energy level of the cathode is localized below the LUMO level of the chains, the electrons generally do not have enough thermal energy to be directly injected (by ion thermoemission mechanism) in the chains at the interface. However, the development of a potential energy gradient allows an effective charge injection by tunnelling effect from the cathode's Fermi level to chains having a similar level for the LUMO within the plastic layer. The same considerations are applied to the *hole*-injection process at the anode. It is worth mentioning that resorting to a tunnelling effect at the cathode is also required by the frequent formation of an insulating layer (characterized by a LUMO level much higher in energy, therefore inaccessible for the injection) between the metal and the organic layer, due to the oxidation of the metal surface or due to chemical reaction between conjugated chains and metal atoms. Then, the exact nature of the interface between the organic materials and the electrodes fundamentally rules the injection process, and consequently the efficiency of the system.

Once injected, the charges become polarons and start to migrate in the plastic layer caused by the action of the electric field created by the potential difference. Because the organic phases are often disorganized, this displacement is done by successive hops of charges between adjoining chains. This transport mechanism leads to a low mobility of charge carriers and often limits the response time of the devices. When a positive polaron meets a negative one in the same chain or adjoining chains, their recombination generates an exciton (i.e. an electronic excitation of the chain), stabilized by attraction Coulombian forces and coupled with a local deformation of the conjugated backbone geometry. According to quantum mechanics, statistically three triplet states and one singlet state are the result of the recombination of the electron and the hole, each of them having a spin of α and β type. However, only singlet excitons contribute to the radiation process in a conjugated material (see below) and can therefore lead to light emission.

These considerations imply that the electroluminescence quantum yield, defined as the ratio of the number of emitted photons over the number of injected electrons, cannot exceed 25 per cent. However, it has been demonstrated that the effective sections of recombination are more important for singlet than triplet excitons. Finally, parts of the photons emitted by the chains have the possibility of leaving the device without being reabsorbed and to this end they go through the anode and the glass substrate.

The triplet electronic excitations generated within the diode are generally turned into heat, since the radiation transition between the triplet state and the fundamental state (i.e. the phosphorescence process) is forbidden by the rule of selection in a lot of organic materials. However, introduction of heavy-metal atoms (for instance, platinum or iridium) within the conjugated backbone or within organometallic complexes allows the efficiency of the process to be improved thanks to relativistic couplings. This is the reason why this strategy is currently developed to allow the simultaneous use of triplet and singlet excitons.

Organic LEDs are a serious challenger to inorganic ones, which can be found everywhere, e.g. in the form of green or red small warning lights on computers and hi-fi systems. The organic LED even currently out performs inorganic ones on many characteristics.

- A potential difference inferior to 5 V is enough to generate quite intense light (between 100 and 200 cd/m^2). This light is moreover emitted in all directions and leads to a high vision angle compared with liquid crystal screens.
- The development of conjugated polymers is particularly easy: they are industrially deposited over an appropriate substrate by spin-coating. In practice, the polymer is first put into a solution (the substitution of the conjugated backbone by alkyl chains makes most conjugated polymers soluble in common solvent) and a drop is then deposited over a spinning substrate, which allows it to spread. Then the solvent evaporates and a highly homogeneous film is formed. This film can spread out over large surfaces (more than 1 m^2), therefore opening up the way to applications inaccessible so far to conventional diodes. This facility of

development is a major advantage compared with inorganic compounds, which necessitate techniques of deposition under vacuum, which are particularly expensive and unwieldy regarding their installation. It is also worth mentioning that LEDs can also be made of small conjugated organic molecules, which also require development under vacuum.

- This technology is more environmentally friendly − 1 g of silicon produces 2 kg of chemical waste.
- The lifetime of devices currently exceeds 80 000 operating hours. This was not easy to achieve given that most of the conjugated materials are sensitive to the presence of water and oxygen (difficult to avoid), which lead to a chemical deterioration process. The solution at the fundamental research level is to work in an airtight box, and at the industrial level the solution is to encapsulate the devices in an organic lattice impermeable to those two harmful elements.
- The great novelty introduced by plastic materials allows the achievement of a flexible screen, foldable without any efficiency lost. This has been demonstrated for the first time by Professor Heeger's group by replacing the glass substrate in a diode by a common polymer and the ITO layer by an electrical conducting polymer. The dominant characteristic of plastic in the device therefore contributes to its flexibility.

A great number of manufacturers in the United States, Europe and Asia are already in the organic-screen market, e.g. Philips, Thomson, Kodak, Dupont, Samsung, Pioneer, among many. May display devices are still at the prototype stage but some of them are already on the market. Most of them are based on a passive matrix. In this case, the electrodes situated on both sides of the plastic layer are respectively formed in rows and columns. The zones where a row crosses a column define pixels of the image that can be independently addressed by applying a potential difference between a certain column and a row. This kind of matrix therefore limits the number of usable pixels because we must be sure that during the addressing cycle the first pixel is still emitting when the last one is reached. A memory-effect mechanism must be added to create bigger screens (such as a computer screen); this is typically realized in active matrices by combining each pixel with a transistor, and many manufacturers are currently working in this area. Therefore, current screens are small devices that could be used for instance in electronic diaries, mobile phones, watches or clocks. Two examples of devices are illustrated in Figure 6.9: (i) the first commercial plastic screen marketed by Philips, used to display the battery level in its new-generation shaver; and (ii) a plastic screen developed by Cambridge Display Technology, a spin-off company initially created by Cambridge University to develop the technology stemming from its discovery. This slimline colour screen (around 1 mm thick) is the size of a mobile-phone screen and contains around 30 000 pixels. The screen is made of pixel rows which emit alternatively the three primary colours (red, green and blue) and have been deposited by an ink-jet-printing-like process with a resolution around a micrometre.

OLED (*organic light-emitting diode*) technology has become common in our daily lives, and will soon be used in lighting domains as well. Here, it is a case of combining many different materials to generate a white light. Recent research works demonstrate

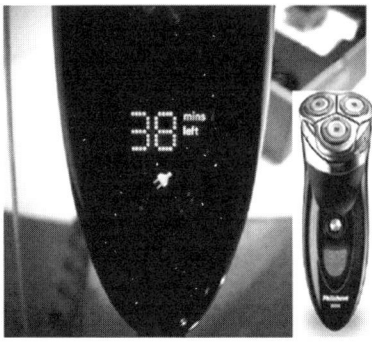

Figure 6.9   Display devices based on conjugated polymers: the Philips low-resolution screen on a shaver (left), and high-resolution screen (around 30 000 pixels for a size of 5 cm on the diagonal) developed by Cambridge Display Technology (CDT)

that organic devices are a great challenge to conventional bulbs; they even out perform them regarding their performances such as the luminous intensity, luminescence yield, lifetime and production cost. As regards flexible screens, they become real products thanks to the Universal Display Corporation: they have recently demonstrated the fabrication of flexible display screens based on small organic molecules. Therefore, with this technology it is conceivable, for instance, to develop pens with a flexible screen. Another field of application, not really currently very advanced, involves developing notice boards or hoardings of large size on any type of support and any size. Philips has even planned to develop luminous wallpapers.

6.5.2   Photovoltaic sensors and organic solar cells

Unlike with LEDs, the functioning principle of a photovoltaic sensor is to convert light into an electrical current; a solar cell transforms solar energy into electrical energy. Unlike diodes, the typical architecture of a photovoltaic cell consists of a glass substrate on top of which is successively deposited a transparent metal electrode, a plastic layer, then a second transparent metal electrode of a different nature. The connection of the two electrodes leads to a levelling of the two metals' Fermi levels and the generation of a potential energy gradient within the organic lattice, which will favour the separation of charges (we will analyse this in depth later). It is worth mentioning that no potential difference is applied between the two electrodes in a solar cell, since its aim is to produce energy without consuming any.

Four different successive mechanisms will rule the conversion of light into electrical current. First, the incident radiation goes through the devices via the transparent surface and is absorbed by the material in the active layer, therefore creating electron-hole pairs. In order to manufacture high-performance solar cells, it is necessary that the material used in the devices absorb radiation on all the solar-emission spectrum that has wavelengths up to 700 nm ($\sim$1.8 eV) in the infrared. This process is far from being optimized for most conjugated materials currently used, which typically absorb

(a)   (b)

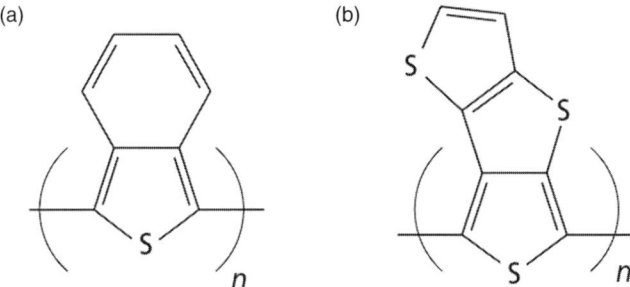

Figure 6.10   Chemical structure of polyisothianaphtene (a) and polydithienothio-phene (b)

energy above 2 eV, and requires the synthesis of new materials, having a low energy gap between their HOMO and LUMO levels. The few conjugated polymers that satisfy these requirements – such as polydithienothiophene and polyisothianaphtene characterized by a forbidden band of around 1 eV – are not currently used because of their lack of stability and ease of development.

Subsequently, the generated electron-hole pair must be separated and transformed into free charge carriers. However, the electrical field generated between the two electrodes as a result of the levelling of Fermi levels is not strong enough to dissociate the excitons because of the high bond energy between the electron-hole pairs within the conjugated material (typically superior to 0.3 eV). This is the reason why a mix of two different materials (one that can easily give π-type electrons and one that can easily receive) is required to dissociate the excitons by a process of photoinduced charge-transfer. This mix can, for instance, be between paraphenylene polyvinyl as donors and C_{60} fullerene molecules as acceptors (Figure 6.11a). If the donor is initially brought in its first excited state by an electronic transition between its HOMO and LUMO levels, then the excited electron can be transferred on the LUMO of the acceptor (with lower energy) while the hole is still located on the donor. Consequently, an exciton is converted into separated charges by a transfer process of photoinduced electrons. On the contrary, if the acceptor is initially excited, a transfer of photoinduced holes will lead to a dissociation process (Figure 6.11c). In a third stage, the generated charges migrate through the plastic layer under the influence of the electric field, in order to be finally collected at the electrodes, to be used in an external circuit.

The two compounds used within photovoltaic cells can be in the form of two different layers within the organic matrix. This arrangement is not the most favourable since only the excitons formed next to the interface between the two materials are likely to contribute to the generation of the electrical current. This zone has a characteristic length of a few dozens of nanometres, corresponding to the mean path (also known as *diffusion length*) of excitons in an organic film. Moreover, this layer-type structure is even more unfavourable because the quantity of absorbed light decreases exponentially with the distance through the film, and therefore is not optimal in the interfacial region. Consequently, a better approach would be to create a homogeneous

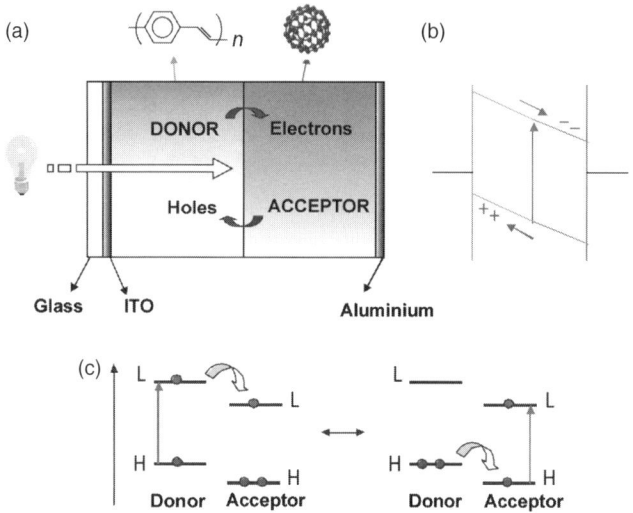

Figure 6.11   *Photovoltaic cell: (a) typical architecture of a photovoltaic cell based on a mixture of paraphenylene polyvinyl and fullerene in contact with two different types of electrode (ITO and aluminium); (b) energy diagram illustrating the light-absorption process and charge migration within the solar cell; (c) illustration of the transfer process of photoinduced electron (left) or of holes (right), which is responsible for the conversion of excitons in charge carriers*

mixing of two compounds at the nanometre scale in order to make access to the interface easier, and increase its total surface. However, this technique is incompatible with the common immiscibility of organic compounds, leading to phase-separation processes. In this context, techniques of local probe microscopy are the best tools in order to characterize, understand and finally control the arrangement of a mixing of two compounds at microscopic scale, as will be described later.

By now, the best power yield (defined as the ratio of the number of charges susceptible to being created by the incident photons and the number of charges effectively collected at the electrodes) obtained for an organic solar cell is around 3 per cent. This value is inferior to the current yield of cells based on amorphous silicon used, for instance, in solar pocket calculators (around 10 per cent), or more expensive devices based on crystalline silicon used in aerospace applications for instance (around 15–20 per cent). Devices based on conjugated polymers do not really compete (yet) with those based on inorganic material regarding performances; however, they distinguish themselves by their low-cost development, flexibility and lightness, and ability to cover very large surfaces. Research in this area has recently led to the fabrication of a flexible solar cell by Professor Serdar Sariciftci's group (University of Linz). This cell is based on PPV and C_{60} with an active surface of 80 cm^2 (Figure 6.12). Very soon, we will find such solar cells on the roofs of houses to supply individual energy

Figure 6.12   Solar cell: flexible solar cell based on PPV and C$_{60}$ with an active surface of 80 cm^2 [Source: Professor Serdar Sariciftci, University of Linz/Johannes Kepler of Linz]

sources or other devices (mobile phone, video camera, radio relays and so forth) for an autonomous electrical supply.

Finally, it should be noted that some transfer processes of photoinduced charges can also be used within chemical sensors (devices that can detect and/or measure quantitatively a given chemical compound). The group of Professor Timothy Swager (Massachusetts Institute of Technology) has recently developed a landmine detector sensitive to vapours of trinitrotoluene (TNT), with a detection threshold of 10^{-15} g. This device involves highly fluorescent conjugated polymer chains subjected to a photoinduced charge-transfer process in the presence of TNT molecules acting as acceptors. Because this transfer dissociates the exciton, and prevents its radiation fallout at the polymer chain level, the detection of targeted molecules is monitored through the evolution of the fluorescence of the material.

6.5.3   Organic transistors and plastic integrated circuits

A transistor is a three-terminal electronic component, in which the passing of the current between two terminals is controlled by the third terminal. In the case of field-effect transistors (FET), it is the potential applied at the 'gate' that gives the intensity of current passing between the 'source' and the 'drain'. By modulating this gate potential, it is possible to set the state of the transistor (conducting or blocking) by changing the conductivity of the material within the device. This process is at the root of the binary data processing, on which are based modern microelectronics and its applications, which go far beyond computer devices. Nowadays, most electrical devices (household appliances, TV, hi-fi, toys, etc.) contain microprocessors based on integrated circuits, a complex assembly of transistors that perform logical operations and data storing.

Microprocessor performance depends directly on the number of transistors each contains. Therefore, the microelectronics industry has invested a great deal to reduce the size of individual transistors and consequently increase the capacity of integrated circuits. This constant evolution is expressed through Moore's law, mentioned earlier (and see Figure 4.1). In parallel with these miniaturization efforts, research work in microelectronics aims for the integration of integrated circuits in new applications such as smart cards and electronic tagging devices for monitoring and identification of manufactured goods. With this new technology, it will be conceivable to walk through a detector gate with the shopping trolley in a supermarket and to have the total price displayed on the cash register almost instantaneously, or even to have a refrigerator displaying the name of foods past their use-by date or exhausted. These devices must be relatively cheap (it is hardly conceivable that a label would cost more than the product it was stuck on) and often require flexible support. Given that common transistors are based on inorganic material such as silicon, they can be expensive to produce and hardly withstand mechanical stress (the support on which they are built cannot be bent). This is why new transistors based on organic materials (polymer or molecular) are being developed, where production costs are low and fabrication on a flexible medium does not cause any difficult problems. Figure 6.13

Figure 6.13   Plastic integrated circuit produced by Philips

shows an integrated circuit containing more than 300 transistors completely made of polymer material, developed by Philips.

6.5.4   Field-effect transistors

Before analysing the material used to build 'organic' transistors, the principle of the FET will be demonstrated once more, underlining the parameters that rule its performance. Figure 6.14 shows a diagram of an FET. During the fabrication of this structure, the 'gate' electrode is first covered by an insulating layer, then the 'source' and 'drain' electrodes are deposited, and finally covered by a thin layer of semiconductor. Actually, only the semiconductor between the two electrodes plays a role in the functioning of the device. It is important to mention that the transistor can be completely fabricated in plastic materials, by using simple and cheap fabrication methods. Therefore, on an initial plastic substrate (e.g. polyethylene terephthalate, PET) can be deposited the 'gate', based on an electrical conducting polymer such as the PEDOT/PSS described earlier, as well as the insulating layer, e.g. a polysiloxane, by spin-coating. Then the 'source' and 'drain' electrodes (based on EDOT/PSS) can be deposited using a process – similar to ink-jet printing – with an aqueous solution of this polymer. Finally, the semiconductor layer is deposited by evaporation under vacuum (if the semiconductor is in molecular compound form) or by 'printing' (from a semiconducting polymer solution).

The semiconductor used is often slightly p-doped, i.e. contains a small quantity of positive mobile charges. Its electrical conductivity is therefore very low and the application of a potential difference between the drain and the source, known as V_{DS}, generates only a very small electrical current; the transistor is then in the state 'off' (blocking state). When a potential difference is applied between the source and the gate (this potential is known as V_G), an electric field is generated through the insulating layer, expressed as an electric-charge accumulation in the semiconductor close to the interface with the insulator (this phenomenon is similar to the charge of a capacitor). If the sign of the charges formed at the interface is the same as the one of the defects already present in the semiconductor, the density of the charge carriers of the semiconductor (therefore, its electrical conductivity) increases when V_G is applied. Consequently, when V_{DS} is applied, a current I_{DS} proportional to V_{DS} is generated as illustrated on the left-hand side of the curve in Figure 6.15. The

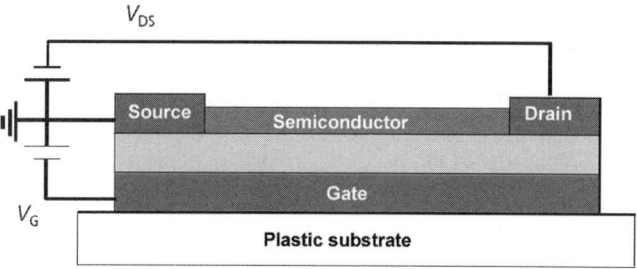

Figure 6.14   Field-effect transistor

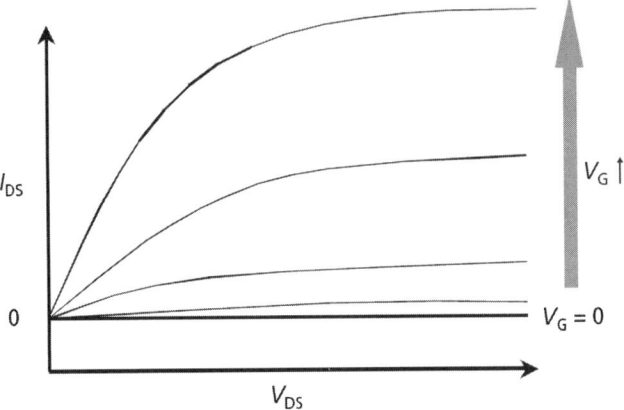

*Figure 6.15   Intensity of the current in an FET. Curves of intensity of the current I_{DS}
going through the semiconducting layer, from source to the drain as a
function of the potential difference V_{DS} between these electrodes, for
increasingly negative value of the potential difference between source
and gate (V_G). The transistor considered is an FET, which contains a
p-type semiconductor; for the negative value of V_G, additional carriers
will be created at the semiconductor–insulator interface*

transistor is then 'on' (passing state). For a given value of V_G, the current I_{DS} reaches
a saturation when V_{DS} increases (right-hand side of the curve in Figure 6.15). The
current is limited by the slow decrease of carriers along the canal. This limitation
can be overcome, to a certain extent, by increasing the potential V_G. The associated
increase in the electric field leads to the generation of a greater number of charge
carriers, i.e. a higher conductivity. Therefore, for a given value of V_{DS}, the current
I_{DS} increases with the potential V_G.

   Recent research work demonstrated that in organic transistors, the transport of
charges is entirely done in the two first molecular layers in contact with the insulating layer. The active layer of the semiconductor is then extremely thin, around a
nanometre. From a fundamental point of view, this is a good illustration of the 'nanotechnology' aspect of these devices. From a practical point of view, this implies that
a very small amount of semiconductor material is enough to make these transistors.

6.5.5   Purity of compounds and field mobility

FETs essentially depend on two parameters: the ratio between the value of the current
in the 'on' and 'off' states and the mobility of the charge carriers. Therefore, in order
for these transistors to be used to control pixels in liquid-crystal flat screens (another
application considered for these devices), the ratio on/off must be of the order of 10^6.
This implies that the conductivity in the blocking state is very low. Therefore, it is
necessary that the chemical compounds be extremely pure (chemical impurity in the
semiconductor layer will favour the passing of the current when V_G is equal to zero).

The current methods of organic synthesis and intensive purification allow to reach a purity above 99 per cent.

The second important parameter is mobility, which corresponds to the drift velocity of charge carriers per electric field unit. For a transistor geometry and a given density of charge carriers, the conductivity and the intensity of the current going through the device are directly related to the mobility of the carrier.

In the saturation regime, the relation linking the current I_{DS} to the mobility μ of the charge carriers is given in (6.2), where W is the width of the canal, L its length (i.e. the distance between the source and the drain), C the capacitance by surface unity of the insulating layer, and V_0 the threshold value of V_G beyond which the transistor becomes passing:

$$I_{DS} = WC\mu(V_G - V_0)^2/2L \tag{6.2}$$

From this equation one can see that the current directly depends on the mobility, but also on the geometry of the transistor. To maximize I_{DS}, the width of the canal can be increased, but this is in contradiction to the objectives of device miniaturization. It is more advisable to work on the diminution of the length of the canal. Therefore, it is currently possible to make organic transistors with an L of a few micrometres by using low-cost techniques such as that similar to ink-jet printing.

In principle, it is also possible to intervene on the capacitance of the insulator, but the properties of the transistor will not change a great deal. However, the mobility in the semiconducting layer can vary over many orders of magnitude, in the function of the chemical structure and the molecule arrangement within the layer. It is therefore essential to improve this property.

For considered applications, the mobility should be at least 10^{-2} cm^2/V.s. In inorganic crystalline semiconductors, such values are easily reached and even exceeded by many orders of magnitude, which allow for fast integrated circuits at the heart of computers (the amorphous silicon has a typical mobility of 10^{-1} cm^2/V.s). However, the mobility is usually much lower in organic semiconductors (to be analysed in depth later); this is why new strategies of synthesis of materials and elaborations of thin films are developed to obtain maximum mobility.

6.5.6 Ideal structures and possible ones

The ideal structure of an organic transistor should consist of an arrangement of conjugated polymer chains (such as PPV), perfectly regular and parallel to each other, connected to the source by one end and to the drain by the other end. The displacement of charges, being easier along these chains (see previous section on electronic structure of polymers), and this structure give outstanding performance to the transistor. Unfortunately, this device will not be developed any time soon, because, on the one hand, conjugated chains long enough to connect one electrode to the other are very hard to synthesize (even if it were feasible, they would contain structural defects that would hinder the displacement of the carriers); and, on the other hand, a perfect alignment between electrodes is almost impossible to make. Therefore, in a real semiconducting polymer layer, charge carriers must perform a great number of 'hops' from one chain to the other during the crossing of the device. These hops are a restrictive factor of carrier displacement, all the more when the molecules are disorganized in the layer (generally the case for polymer chains). This is why the mobility of semiconductor polymers is generally low, around 10^{-4}–10^{-5} cm^2/V.s.

At the microscopic scale, the hop of a charge carrier from one chain to another corresponds to the transfer of a polaron of the initial chain to the other chain, which was initially neutral. The electronic levels involved in this process are the orbitals at the boundary of the two π-type chains (i.e. HOMO levels for the holes' displacement, and LUMO for the electrons' displacement). Therefore, the transfer will be more efficient if the orbitals at the boundary of adjacent chains strongly interact. That is to say, the electronic density of the two chains must widely overlap, implying that the chains are close and parallel to each other. To favour this compact stack of regular and parallel chains, it is necessary that the chemical structure be regular. It is the case for instance for the stereoregular poly(3-alkylthiophenes), in which the alkyl group (represented by R in Figure 6.16) is attached to the same carbon atom on every thiophenes ring. The mobility in a highly organized thin film of this polymer is of the order of 10^{-1} cm^2/V.s, far greater than the figures obtained for the same polymers but non-stereoregular (Figure 6.16).

A highly organized arrangement is rarely developed over long distances in polymer materials, particularly because of the distribution of the molecular masses (all chains do not have the same length, which disturbs the organization of a regular lattice within the solid). An interesting alternative consists of using conjugate oligomers, i.e. molecules corresponding to short conjugated polymer segments; moreover, such compounds can be synthesized with a great purity. This is the case for instance for the sexithienyl, the bis(dithienothiophene) or the pentacene, chemical structures illustrated in Figure 6.17. Like the conjugated polymers, these compounds have a π-type electrons system delocalized along the molecule. Furthermore, the chemical purity of these compounds makes them able to form high crystalline thin films. The combination of these two advantages (π-type orbitals at the boundary and well-organized arrangement over long distance) gives these materials high mobility for charge carriers (up to 1 cm^2/V.s at ambient temperature). In order that this high mobility can

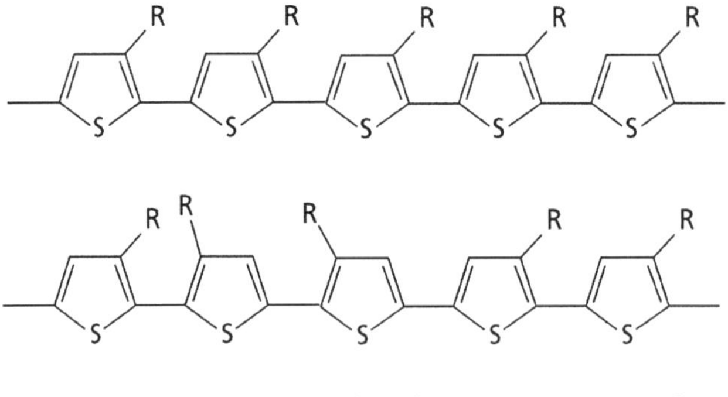

Figure 6.16   Chemical   structure   of   chain   segments   of   stereoregular poly(3-alkylthiophenes) (above), and non-stereoregular (below)

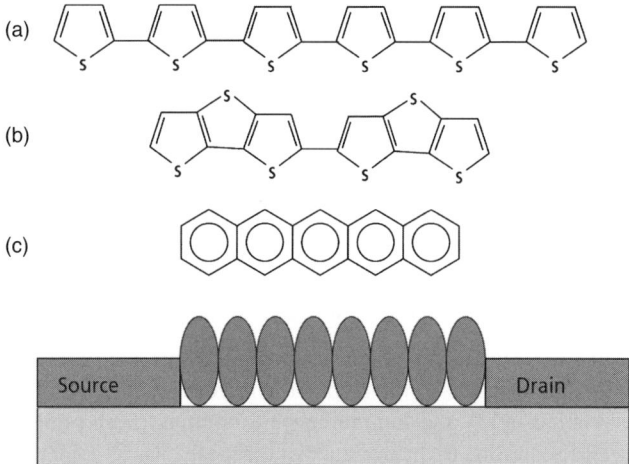

Figure 6.17   Chemical structures of sexithienyl (a), bis(dithienothiophene) (b), and pentacene (c). Diagram of the optimum molecule arrangement in the canal of the FET. The molecule are all oriented perpendicularly to the direction source → drain, and their π-type electronic densities overlap, in order to favour hops of the charge carriers

be expressed by the easy crossing of the current from the source to the drain, it is necessary that the molecules in the first layers be oriented perpendicularly to the insulating substrate surface (as illustrated at the bottom of Figure 6.17). It is the preparation of the insulating surface, associated with the chosen method of depositing the organic semiconductor, that favours the development of the semiconducting layer according the appropriate molecular orientation.

6.5.7   Polymer-based biochemical sensors

Sensors are measuring devices that are able to selectively recognize chemical or biological molecules, and convert this identification into an electrical or optical signal. Moreover, the amplitude of the detected signal is usually proportional to its concentration, allowing the quantitative proportioning of these species from a calibration curve. The advantages of such an 'artificial nose' are numerous.

- Chemical sensors can be used to detect 'hazardous' molecules in fields such as food safety and the environment. They also have high sensitivity, which allows unambiguous and fast identification of chemical substances present in low quantities, e.g. vapours of explosive within the context of antipersonnel mine detection.
- Biosensors are invaluable tools for medical diagnostics or genetic mutation (DNA and proteins detection). According to the World Health Organization (WHO), the development of a low-cost and reliable method for identifying infectious pathological agents is one of the main critical steps for the improvement of health

levels in developing countries. They present many other potential uses, including dosage of glucose in blood (detection of saccharose) or analysis of enzymatic activity (enzymes detection).

Sensor principles imply two different stages: (i) the selective recognition of molecules to measure out via specific interactions with the active constituents of the sensors; and (ii) the expression of this process of detection as a signal that is easy to detect and quantize using common analytical methods.

Therefore, the devices are composed of two connected elements: a sensing element and a transducer. Sensors based on conjugated polymers can be classified according to type of measurement used during this transduction step.

- **Optical transduction**: In this case, the conjugated polymers are potentially interesting candidates because of their optical characteristics (high absorption coefficient for visible light) and luminescence properties (high-photoluminescence quantum yield according to the presence of the molecules to detect). The principle of the sensor is based on the selective development of complexes of polymer-based molecules and the resulting change in the photoluminescence spectrum. The amplitude or the wavelength of the luminescence signal can vary in the presence of analytes (substances or chemical constituents that are determined in an analytical procedure, such as a titration).
- **Electrical transduction**: This method uses the modification of the properties of charge transport within the conjugated polymers (neutral or doped) in the presence of the analyte. The presence of small molecules can affect the mobility (and even the number) of charge carriers in the polymer active lattice. The measurement of the electrical current (as a function of the tension applied to a device such as a transistor in the case of a semiconducting polymer) or the electrical conductivity (in a device including a chemically doped polymer layer) allows detection of the compound.

6.6   Photoluminescent conjugated polymers

Sensors based on luminescent conjugated polymers are at the centre of a lot of research because of an ability to detect analytes in very low concentrations. The cancellation or existence of the luminescence in the presence of molecules to be sensed is effective as a form of analysis because of its high sensitivity and the simplicity of the 'on/off' detection method. Sensors where the development of a complex between the conjugated polymer and the species to sense leads to a spectral shift of the light emission can also be designed. The advantage of this approach is that titration can be based on the ratio of the optical intensities of the two wavelengths rather than being based on their absolute value, which avoids possible photo-degradation problems (in that case, the overall intensity decreases without affecting the ratio). Devices based on this method are known as *ratio-metric probes*.

Before discussing possible applications of luminescent conjugated polymers for the correct proportioning of chemical or biochemical species, it is useful to analyse

(a) molecular-based sensor

(b) polymer-based sensor

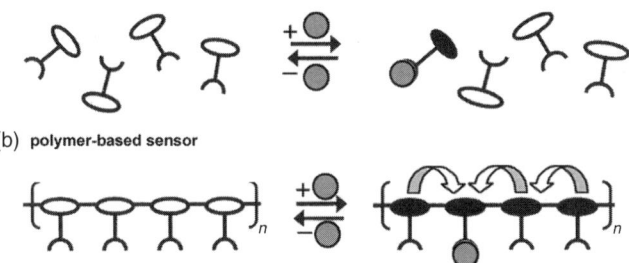

Figure 6.18   Molecular-based vs. polymer-based sensor. Illustration of the working principle of a molecular-based sensor (a) and a polymer-based one (b). In the case of a molecular-based sensor, the insertion of an analyte (represented by a grey disc) leads to a state shift of a single molecule from bright (represented by a white oval) to dark (represented by a black oval). In the case of the polymer, the insertion of an analyte leads to the simultaneous state shift of a great number of monomer units connected by conjugated covalent bonds. [Source: adapted from J. S. Yang, T. M. Swager, Journal of the American Chemical Society, *120, 5321 (1998)]*

the origin of the high sensitivity of polymers. It has been demonstrated that the detection threshold can be lowered via the use of conjugated polymers rather than using molecules of similar chemical structures. The boost of the sensor response in polymers is due to their delocalized electronic structure. Even if this delocalization is interrupted by many structural and/or chemical defects (as mentioned above), the chromophores (part of the molecule responsible for its colour) stretch out over many monomer units, and, moreover, they are coupled with each other via electronic interactions. Then the different repetitive units of the polymer behave in a collective way – at least in an interval of time of the order of 100 fs after photo-excitation – which gives a synergy to the optical response of the material. This concept is illustrated in Figure 6.18, which compares the working principles of a polymer-based sensor with a molecular one. While in the molecular approach, the development or the disappearance of an analyte leads to a shift in the luminescence state (bright or dark) of a single molecule, the electronic state of a great number of repetitive units is affected by the presence or not of the analyte for a polymer. It is as if the excited species created by photon absorption explore a great number of receptor sites along the polymer chains during their lifetimes, which leads to an amplified response of the sensor compared with the response of molecular systems.

6.6.1   Chemical sensors

As briefly described above, the inhibition of the luminescence in the presence of chemical species can be used to detect these species with high sensitivities due to a synergy effect between repetitive units of polymer chains. The selectivity regarding a species or a group of species is another essential parameter within the specifications of a good sensor. Using the richness and flexibility of the organic synthesis, researchers have developed many strategies in order to define chemical structures of polymers likely to be used as selective receptors of some group of molecules. It is obvious that the development of a complex between the conjugated material and the substance to detect must be reversible so that the sensor may be restored to its initial state after a detection process. Many developments of such chemical sensors have been described in scientific literature. Below, the working principle of a TNT sensor is analysed in depth.

Metal detectors are normally used to locate antipersonnel mines, an exercise that can be relatively expensive. The use of chemical sensors to detect the vapours of TNT contained in mines therefore appears very attractive. Timothy Swager (MIT, Boston) has developed a polymer-based sensor capable of detecting very small quantities of TNT (in the order of the femtogram, 10^{-15} g) in a few seconds. The active component of the sensor is a conjugated polymer, derived from the poly-*p*-phenyleneethynylene also known as PPE and made functional by pentiptycene lateral groups (Figure 6.19). This 'functionalization' of the polymer plays an essential role because it leads to porous films that allow the diffusion of vapours of TNT within the film and the development of complexes between conjugated polymers (rich in π-type electrons) and TNT molecules (poor in π-type electrons). The selectivity regarding the explosive is ensured by geometric as well as electronic criteria. TNT molecules have the appropriate size to be inserted between the polymer chains, and are also good electron

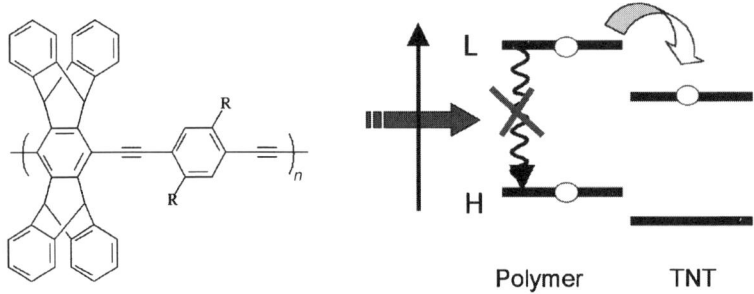

Figure 6.19   *Chemical structure and working principle of a polymer-based trinitro-toluene sensor. TNT molecules in the gaseous state diffuse within the polymer films, and form, with the polymer chains, stabilized complexes by π–π interactions in the pentiptycene units. The optical excitation of the polymer material (creation of an electron-hole pair, in the right-hand side of the schematic) is followed by an electron transfer (white arrow) from the polymer LUMO level to the LUMO level of the TNT molecule (which is a very good acceptor of electrons). This process competes with the radiation fallout on the polymer (zigzag arrow); the intensity of emitting light of the polymer is then reduced in the presence of TNT*

acceptors (due to the presence of nitro groups). In the presence of TNT, the photoluminescence of the conjugated polymer disappears as a result of a transfer of electrons in the photoexcited polymer chains towards the TNT molecules. This process leads to the development of a species whose charge transfer is non-emissive (with the hole being on the polymer chain and the electron on a TNT molecule) (Figure 6.19). Nomadics Inc. has developed a portable detector of antipersonnel mines based on this technology.

6.6.2 Biological sensors

The development of fluorescent polymer-based sensors for the detection of biological molecules offers many development opportunities in the field of medical diagnosis. Shorter analysis times, higher sensitivity and minimum margins of error open up the way to more accurate and cheaper diagnosis methods. Below, an example of the use of conjugated polymers for the identification of DNA is analysed.

Since the discovery of its double-helix structure, recognition properties of DNA strands via hybridization reactions between complementary bases have been well known. But to identify the appropriate transduction processes that are likely able to generate a measurable signal resulting from these identification stages is another matter. Professor Leclerc and his group have recently proposed a novel approach that allows the detection of only just a few hundred DNA molecules by means of a biosensor based on a cationic polythiophene, which is fluorescent and soluble in water. Like all polyelectrolytes, positively charged polythiophenes form stoichiometric complexes with negatively charged oligonucleotides (Figure 6.20). These neutral complexes (or 'duplexes') precipitate in solution and the resulting aggregates generate a moderate light that shifts in the red range (compared with isolated polymers). However, in the presence of complementary DNA strands, the photoluminescence intensity increases and shifts towards blue. The hypothesis suggested by researchers to explain this luminescent change is based on the winding of polymer chains around two strands formed by hybridization between the complementary segments. The structures formed in this way (known as *triplexes*) are more soluble and are characterized by a photoluminescence spectrum very different from the one of the aggregates. The mechanism is not well understood; however, this approach allows for the detection of DNA at very low concentration close to the zeptomole per litre (10^{-18} mole/l), and outstandingly allows different structures to be distinguished by only one nucleic acid. The method has been successfully tested to identify the type-A influenza virus in a flu diagnosis test (Figure 6.21).

6.7 I/V response of field-effect organic transistor

The sensors described above are based on a shift of the luminescence or a change in the absorption properties of conjugated polymers due to the detection of an analyte. Another quantity that can be used as a sensorial response is the electrical current. In this context, a novel approach currently in a development stage consists in measuring

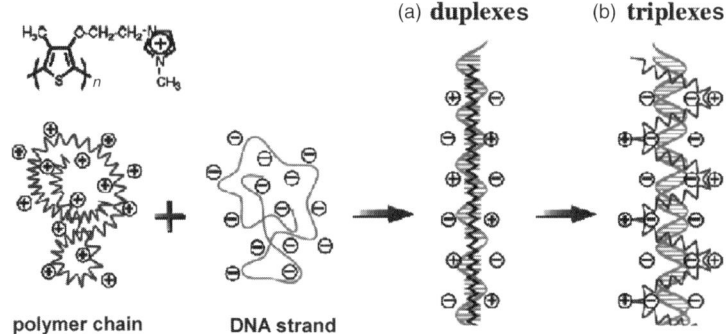

Figure 6.20   *Working principle of a DNA sensor based on conjugated polymers.*
(a) 'Duplexes' are structures stabilized by electrostatic interactions
between negatively charged DNA mono-strands and positively charged
polymer chains. These structures aggregate in solution, leading to a
moderate light emission in the red range. (b) In the presence of comple-
mentary DNA strands, double strands are formed by hybridization. The
polymer chains wind around the DNA double helix, forming 'triplexes',
structures that are more soluble and are characterized by a more intense
light emission and shift in the blue range. [Source: K. Dore, S. Dubus,
H.A. Ho et al., Journal of the American Chemical Society, *126, 4240*
(2004)]

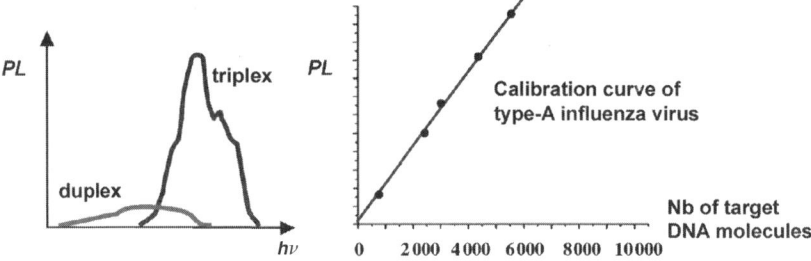

Figure 6.21   *(a) Illustration of the shift in the photoluminescence (PL) spectrum*
from the 'duplex' structure to the 'triplex' one; (b) calibration curve
obtained for type-A influenza virus. The detection threshold is around
320 molecules in 150 μl i.e. around 10–18 mole/litre [Source: ibid.]

the electrical variation (due to the presence of the species to detect) in a transistor,
which involves a conjugated polymer in its active layer. Transistors are the main
elements of any electronic circuits used to process or to store data, like the memory or
the processors of computers. The most commonly used transistors are in silicon. Over
the last few years, the development of organic electronics has led to the fabrication
of transistors based on conjugated organic materials (polymers or small molecules),
as described below.

The use of a transistor as a sensor is based on a change in the electrical characteristics in the presence of an analyte. Doctor Ananth Dodabalapur's group have demonstrated that, in transistors based on small conjugated molecules, the current I_{SD} varies up to 25 per cent when exposed to vapours of strange molecules. Mechanisms likely to be responsible for this current variation are numerous and not currently well understood. On the one hand, the presence of the analyte can make the current increase by triggering a reduction/oxidation reaction that will introduce additional charges on the conjugated molecules. On the other hand, the current decreases when the analyte seeps through structural defects of the organic layer and disturbs the charge transfer. It has been clearly demonstrated that the morphology of the organic film highly influences the sensor response; highly crystalline structures lead to poorer responses. In most cases, the sensor can be reused many times by applying an important potential difference of an inverse polarity of the one initially applied between the source and the drain, and/or by heating the device. It is worth mentioning that it would be conceivable, with such a low-cost technology, to bin the sensor after use. The main problem with these devices is their lack of selectivity regarding the species to detect; this is why it would be necessary in the future to develop devices (artificial noses) that combine several transistors so that each of them uses a different organic material in order to multiply the sensor responses and allow the determination of target species by crossing results. By exposing a part of the organic layer to a solvent (for instance, an aqueous solution) while protecting the rest of the device using an appropriate impermeable layer, it is also possible to detect a species in solution using transistors. It is finally worth mentioning that organic transistors are very good pressure sensors, as recently demonstrated by the group of Takao Someya at Tokyo University.

6.8   Doped conjugated polymers

Another common method for the detection of chemical species in gaseous phase or in solution is based on changes in the electrical conductivity of a doped conjugated polymer. Conducting polymers such as polypyrrole or polyaniline have been used for stability; they distinguish themselves from metal oxide previously used because of their relatively easy development and low production costs. The electrical conductivity of a polymer depends on the density and mobility of charge carriers; the sensor response can then be based on one of these parameters or both of them together. The chemical species to detect can reduce or increase the charge's density as a result of oxido-reduction reactions with the conducting polymer; it has been demonstrated, for instance, that the electrical conductivity of a conjugated polymer oxidized by agents such as I_2 and Br_2 can be reduced by exposing the polymer to reducing agents such as NH_3 and H_2. An exchange of counterions between a conducting polymer and a solution or the penetration of a gas within a polymer film can lead to a change in arrangement of the polymer chains, e.g. from a globular to an expanse structure, as well as swelling the film, which are both likely to modify the electrical conductivity of the device without changing the density of charge carriers. Such observations

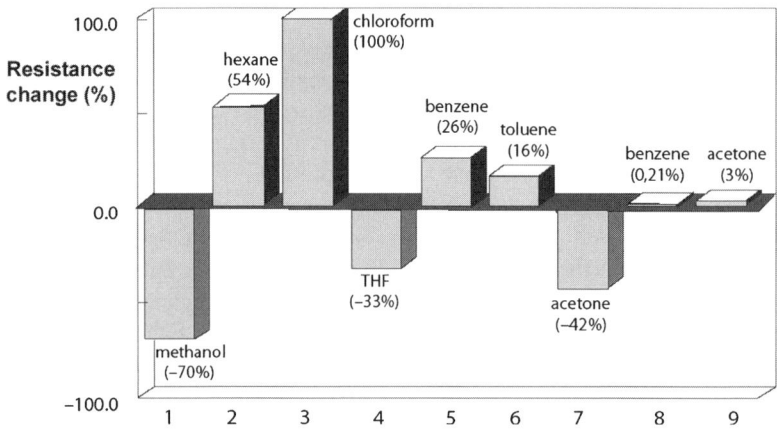

Figure 6.22   Response of a sensor based on conjugated polymer: change of resis-
tance due to the exposure of doped polypyrole films to nitrogen vapours
saturated by different analytes [Source: A.G. McDiarmid, Synth. Met.,
84, 27 (1997)]

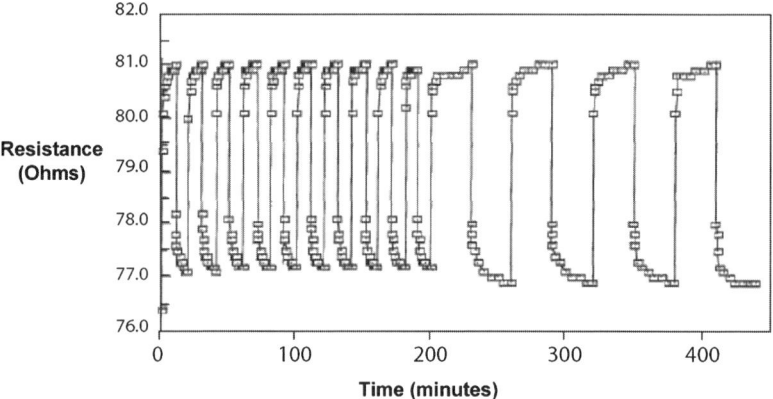

Figure 6.23   Reversibility of the detection process: evolution of the resistance during
cycles of exposure to nitrogen vapours saturated or not with toluene
[Source: ibid.]

have been made for instance by the group of Dr Alan G. McDiarmid by exposing a
film – which contains polystyrene and polypyrole chains doped with dodecylbenzene
sulphonate ions – to vapours of nitrogen saturated by different analytes. Figure 6.22
shows the evolution of the film resistance after an exposure of five seconds (the two
last results have been obtained in the presence of molybdate ions, thus showing the
sensitivity of the response to the nature of the counterion).

Figure 6.23 demonstrates the reversibility of the recognition process; the signals
here were obtained for a sensor based on polypyrrole chains doped by dodecylbenzene

sulphonate ions by exposing the film alternatively to a flow of nitrogen with or without toluene. The general lack of selectivity regarding the species to detect requires the development of a sensor network in order to get accurate results. Each detecting cell of the network gives a different response according its polymer and/or counterion nature, the development condition of the films, and the composition when a mixing based on a conducting polymer and a saturated polymer is used – these kinds of device are developed for instance by Smiths Detection. This technology is likely to have many applications, for instance in the medical, food and environment fields.

Chapter 7
Fabrication of nanostructures

Nanotechnologies are generally defined as the development and use of materials, instruments, and systems at dimensions of the order of 1 to 100 nanometres, i.e. the size of atoms, molecules and supramolecular structures. Strictly speaking, nanotechnologies are revolutionary. At the scale of the nanometre, materials and systems can display characteristics that are completely new, changing not only their properties, but the physical, chemical and biological processes involved as well The changes are so fundamental that properties of the matter at the nanometric scale cannot be deduced from properties at a larger scale. Researchers must be able to control and handle the matter, to characterise and understand the way of using its properties in order to develop new useful devices.

It is not only about science fiction: the microtechnology quest of miniaturization (of the order of the micrometre) is already a reality. For instance, since 1995, at the Todai University in Tokyo, the smallest automatons known have been developed. These are micromachines fitted with six leg antennae, two of them mobile, allowing the micromachines to move around within an electromagnetic field.

Eric Drexler, at beginning of 1990, was the first to follow up Richard Feynman's intuitions. Today the director of the education centre at the Foresight Institute, Drexler is so convinced that nanorobots could one day repair human tissues that he proposed to cryogenically freeze his body after death in order one day to be able to repair and bring it back to life using nanorobots. His credo is simple: if matter can be moved atom by atom, then it is possible to create nanorobots smaller than a virus and fitted with a nanoscopic mechanical arm. These 'assemblers', as they are named, can then build other nanorobots or bigger structures by stacking up atoms and molecules like Lego bricks, thanks to a construction programme that can give instructions simultaneously to millions of nanorobots.

To devise these assemblers, Drexler observed how nature built living organisms. The model he chose is ribosomes. Measuring hardly a few thousand cubic nanometres, these mini-factories are in charge of the synthesis of all the proteins – and consequently all living organisms – on Earth.

There are obviously profound complexity differences between 'life' chemistry and classical chemistry, but nanotechnologists have not given up hope of finding the bridge between these two fields. 'Imagine what our world could be like if we could build, without water and without living cells, devices having atomic perfections as great as the one of living organisms,' wrote nanotechnologist Richard E. Smalley (Nobel Prize in Chemistry 1996).

7.1   Situation of the problem

The major problem of this visionary approach is the control of the position: the atoms must be assembled with (nanoscopic) precision. Therefore, a single hair falling on a nanorobot could have approximately the same effect as an elephant crushing your foot. Moreover, the slightest vibration would prevent any atom manipulation and would lead to side effects and chemical reactions. Additionally, many other problems would need to be solved: what energy would supply these nanorobots; how would the programming and data storage be performed?

Nanotechnologies are a real technological revolution, given that they will theoretically allow mankind to control matter at the molecular level. An examination of the nanotechnology research fields allows a few acknowledgements. The application possibilities of nanomaterials are exceptionally numerous and many industrial sectors could benefit from them.

As described in Chapter 4, computer science was the first field to make use of nanotechnologies. During 1960, the electronics industry adopted the miniaturization approach. Over the years, techniques for manufacturing electronic circuitry became more and more sophisticated. For instance, more than 6 million transistors can be placed on a surface area of 306 mm^2 in a Pentium. In 25 years, the computing power of microprocessors has been multiplied by 25 000. In 15 years, with 10 times more transistors on a few mm^2, a single computer will have the computing power of all the computers now present in Silicon Valley.

But the rush to miniaturization and computing power will soon come up against the problem of 'size': at 100 nm, common lithography processes reach their limit, since the photon wavelength of light becomes too short to etch so thin a pattern in silicon. Researchers reckon that by 2015 only a profound change in the way semiconductors are manufactured will allow the continuation of the miniaturization of electronic components. Moreover, at this scale, quantum effects start appearing; the matter could be in different states preventing the 'on/off' binary logic that rules microprocessors.

With an electron microscope, it is possible to etch lines of 20 to 100 nm in silicon. Smaller than the wavelength of photons, electron wavelengths allow a finer resolution. Instead of etching in silicon, other researchers try to deposit on it nanoscopic circuits, atom by atom or molecule by molecule. We will analyse later in this chapter how supramolecular chemistry can offer an attractive and low-cost solution.

The first tools for the handling of atoms and molecules already exist: in 1986, two Swiss researchers, Gerd Binnig and Heinrich Rohrer, received the Nobel Prize in Physics for their invention, the scanning tunnelling microscope (STM), which allows a view of atoms one by one (the principle of the STM is described in Chapter 4 – see 4.1.2). The STM can detect atoms and molecules, but it can also move them. Groups of researchers at IBM have manipulated atoms in order to write the initials IBM with 35 xenon atoms; they have also built a circle made of iron atoms on a copper surface, thus trapping a free electron in the circle. This quantum box is probably the first step towards the nanoscopic transistor. These first experiments were conducted at very low temperatures (-270 °C), but currently, researchers are able to handle organic molecules of 1.5 nm at ambient temperatures. Recently, they have created the first molecular machine: a Chinese abacus of less than one nanometre, composed of a dozen rows of fullerene molecules (C_{60}) deposited on a copper surface.

Another tool of interest to IBM researchers is the atomic-force microscope (AFM), which detects atoms by measuring the force between tip and surface. On contact with ambient air, a tiny drop of water settles on the AFM tip. By inducing a current, an oxidation is created in the tiny drop; oxide dots of 10 to 20 nm can then be drawn on silicon or a metal surface.

By combining micromotors with optical, mechanical and electronic elements, micromachine manufacturers open the way to a technological revolution as important as the microelectronic revolution: the MEMS or micro-electro-mechanical systems (see Chapter 1).

7.2 Contribution of supramolecular chemistry

Beyond molecular chemistry, based on the bonds between atoms to build the molecule, there is a field of chemistry known as *supramolecular*: the chemistry of associations of two or more chemical species. Supramolecular species are not built using covalent bonds (intramolecular bonds that connect the atoms forming the molecules), but are built from intermolecular interactions. Therefore, it opens up new horizons for the development of nanostructures based on assembling molecular elements, according to the *bottom-up* approach (see Chapter 1).

The handling of atoms using STM to form complex structures corresponds to the bottom-up approach; however, it is essential to mention that supramolecular structures assemble themselves (self-assembling) without any external intervention (such as the use of an STM). The design and the synthesis of molecular constituents that will self-assemble require great skills and creativity from researchers.

The forces that make species combine together are of various types. Fundamental and physical chemistry are applied to specify the properties of these forces, particularly their variation in function, of the nature, the distance and orientation of the different groups involved. Since these forces are generally far lower than that of the covalent bonds, the intermolecular structures can be built and dismantled at a speed depending on their nature, number and position. The development of supramolecular nanostructures requires a real knowledge of the energy and stereochemical aspects of molecular interactions, and particularly an intermolecular conformational analysis, since the achieving of a given structure is based on the fine control of the interactions involved in the bonding of the constituents. Such association is characterized by its stability and selectivity, i.e. by the quantity of energy and information involved; the complex species has its own structure and reactivity.

Self-assembling processes play an important role in biological systems, such as the development of lipid layers, protein withdrawal or interactions between DNA and histones, which play the major role in the regulation of genes' expression. This is how the substrate bonds to the enzyme and hormone to receptor; this is how protein units assemble to form the haemoglobin and multi-enzyme complexes; this is how the double-helix DNA (see 5.2 in Chapter 5) remains in place, and how the genetic code is transmitted via the alphabet of nucleic bases. These systems lie within the scope of the supramolecular arrangement in solution, which has always been one of the major targets of supramolecular chemistry.

At the beginning of the 1970s, it was a matter of forming complexes between small and large entities, the smallest being inserted into the largest. It was the time of 'host–guest' chemistry. The nature of complexed/complexing substrates has since evolved and become more complex; mono- or poly-charged organic ions, then neutral molecules have been able to become even more complex. These self-arranged structures became progressively larger. Currently, as with the process of developing the quaternary structures of proteins, it is often a matter of building self-organized structures of nanometric dimensions involving the self-assembling of units of the same size. Until now, the interactions involved were essentially of short distance and the non-aqueous solvents were favoured. This is why there are many works based on the use of hydrogen bonds, of coordination bonds or of complexes based on charge transfers. Rarer are the works based on hydrophobic interactions, on excluded-volume interaction involving steric effects or on electrostatic interaction.

Another domain of supramolecular chemistry is the one of associations between polymolecular micelles, layers and membranes. The manipulation of molecular assemblies and the physicochemistry of collective systems combined with the chemistry of receptors, catalysts and carriers, open the way to the development of molecular components where photons, electrons and ions are involved. Therefore, a new chemistry can be considered, a chemistry of 'informed' and organized systems, with properties of storing and transferring data, control and calculation at the molecular level. This 'chemistronics' – including photonics, electronics, and molecular ionics – requires deep knowledge of the interactions that rule molecular associations, their transformations.

7.3   Semi-conducting nanoribbons

We have seen that electronics at the molecular scale is an area of research that brings together expectations beyond what can really be achieved. Today, the advances in research at the nanometric scale reduce the gap between dream and reality and mark out new paths in physics and chemistry.

From the fundamental point of view, this field of research is the extension of the current mesoscopic physics to electronic systems in which electrons are quantumly coherent, and on top of that are distributed on atom assemblies as well controlled as molecules. Monolayer carbon nanotubes are currently one of the most striking examples of that kind of system, but supramolecular chemistry and biochemistry allow us to build such rich and innovative systems.

The development of this domain of research is justified by the understanding of the nature of the transport (and/or transfer) of charges and their related effects, in a single molecule or even in an assembly of molecules. The electronic transport phenomena of matter physics are deeply related to the charge transfers involved in chemical reactions. This is why this research is 'viable' if there is a close collaboration between physicists of condensed matter on the one hand, and chemists and electrochemists on the other hand.

To create supramolecular devices for molecular electronics, it is advisable to obtain them at the solid state. Generally, properties of condensed matter are also the result of molecular interactions. This is why the structures and dynamics in liquids, micelles, liquid crystals, emulsions and organized phases use the same kinds of interactions. At the beginning of this chapter, it was explained how molecules could be used as transistors or logic gates of nanometric size. In this section, we will present a methodology that allows the creation, by self-assembling, of one-dimensional nanodevices known as *nanoribbons*. These devices have a typical section of 5 to 20 nm and are an essential step before the creation and control of electronic devices fitted in a single molecule.

Among the conceivable polymers for the achievement of these nanoribbons, conjugated polymers – whose electrical and optical properties were described in Chapter 6 – are particularly of interest. It is worth mentioning that they are characterized by a rigid backbone along which π-type electrons are easily delocalized, thus giving some conducting electrical properties to the polymer. These polymers include poly-para-phenylene (PPP), poly-para-phenylene-ethynylene (PPE), poly-fluorene (PF), poly-indenofluorene (PIF), and poly-thiophene (PTh). Figure 7.1 illustrates the chemical structure of these polymers. The presence of alkyl chains is noticeable; their role is exclusively to make the polymer soluble in most common organic solvents, since generally non-substituted conjugated polymers are insoluble and infusible.

When a small amount of a diluted solution of one of these polymers is deposited on a substrate such as mica or on a silicon wafer, the molecules self-assemble and form nanoribbons of only a few nanometres. This figure corresponds to the length of the molecules. Figure 7.2 illustrates some pictures obtained by atomic-force microscopy

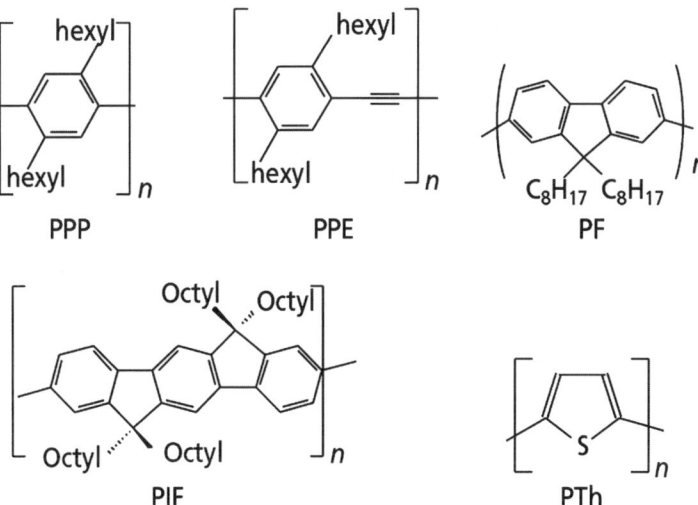

Figure 7.1   Chemical structure of PPP, PPE, PF, PIF, and PTh

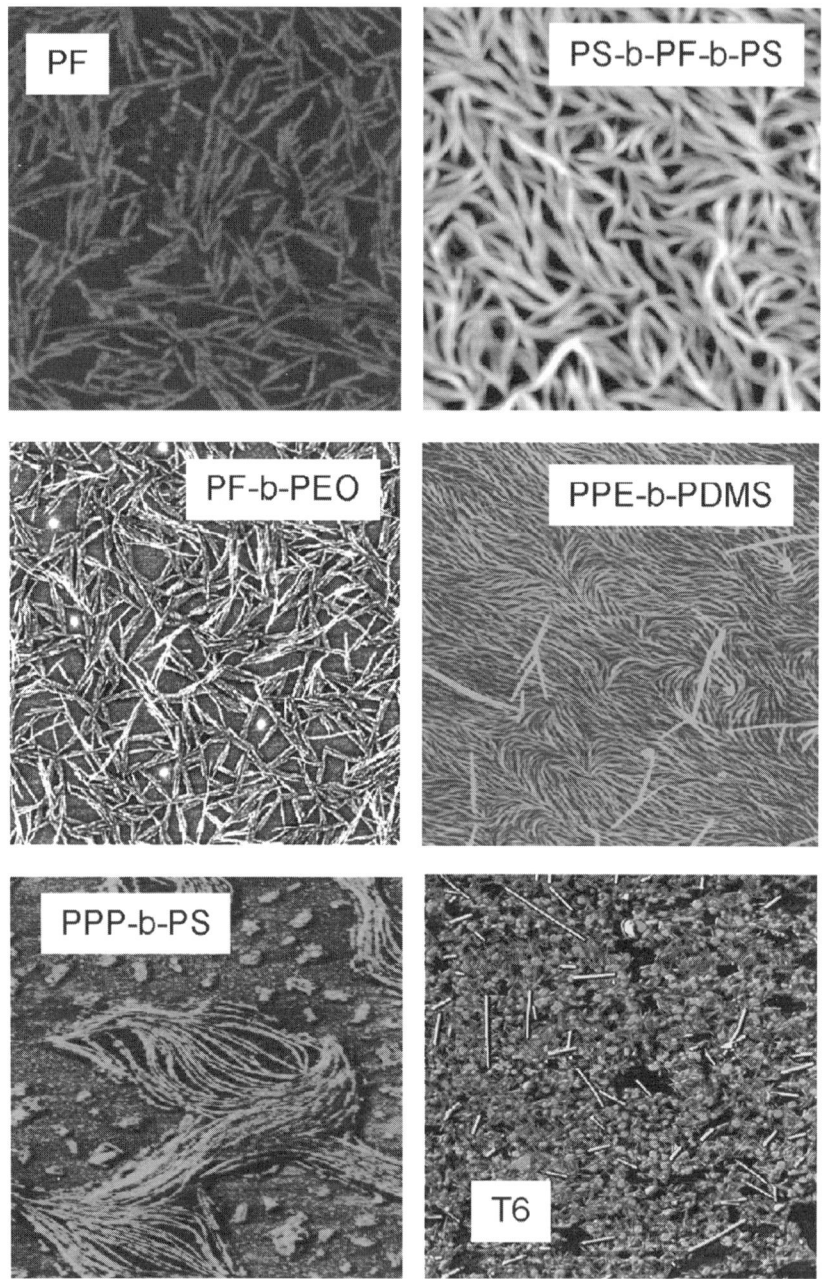

Figure 7.2   *AFM pictures (2 × 2 μm²) of some nanoribbons obtained from conjugated oligomers*

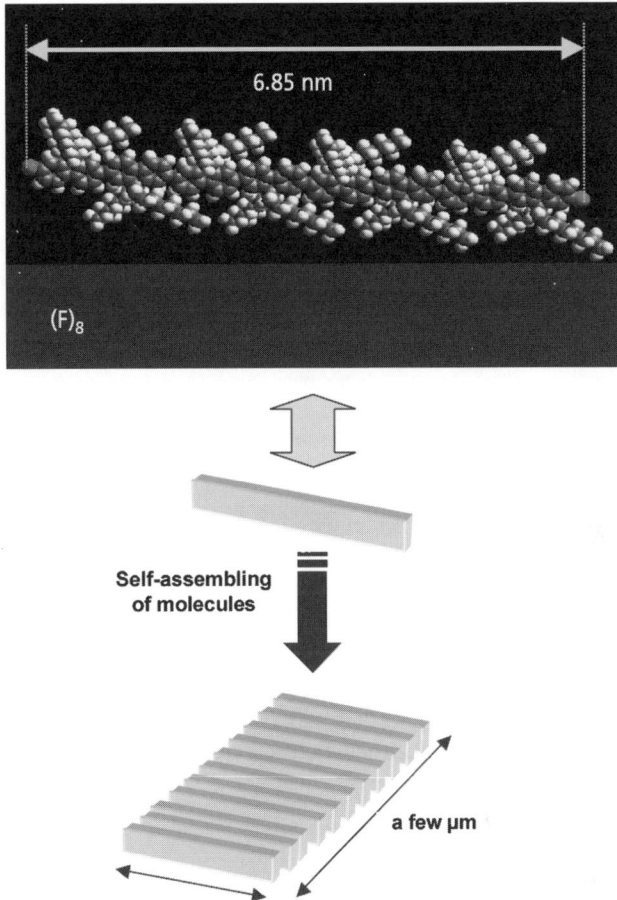

Figure 7.3   Supramolecular arrangement within a nanoribbon composed of fluorine monomers

for different systems having a conjugated oligomer and for some devices having a segment of non-conjugated polymer. Some computation using mechanics and molecular dynamics allows us better to understand the way that these molecules combine within the nanoribbons. Figure 7.3 illustrates the structure of a fluorene oligomer containing eight monomers; the molecules are stacked in parallel like dominoes, and form very long and thin filaments.

7.4   Creation of nanostructures

We have seen how it is possible to create nanodevices from conjugated polymers. Once these ribbons are created, it would be interesting to handle them in order, for

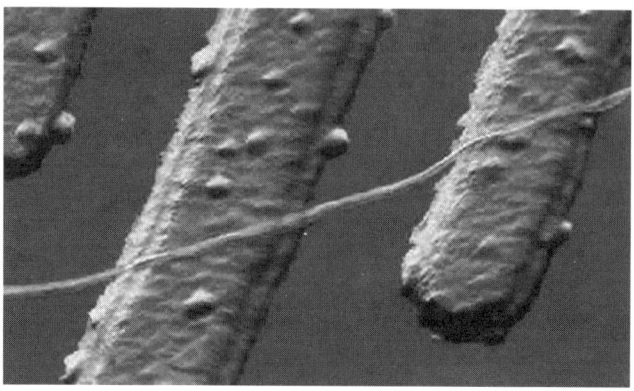

*Figure 7.4   Device composed of a carbon nanotube deposited on two electrodes
[Source: Molecules Biophysics Group, Delft University of Technology]*

instance, placing them between two electrodes (like the works of C. Dekker in the Netherlands, where the conductivity along a carbon nanotube has been measured – see Figure 7.4), to align them in parallel and thus making a complex set of circuits at the submicronic scale. This is known as 'supramolecular electronics'.

A way to make these devices involves forcing the molecules to self-assemble in given directions or predefined regions. It is mostly the nature of the substrate or its structuring (natural or imposed) that will guide this self-assembling. In the next section the different strategies proposed by researchers in order to solve this essential technological problem will be analysed.

7.4.1   Natural structured surface

In nature there are materials whose surface presents:

- domains with different physicochemical characteristics at scales of a few micrometres or dozens of nanometres;
- defects such as holes or bumps;
- particular axes of symmetry;
- an anisotropy of the physicochemical properties, obtained naturally or by rubbing with velvet.

This is why, when a PPE solution is deposited on graphite (Figure 7.5), the fibrils described earlier align themselves according to three main directions. The surface of graphite (of the HOPG type, highly ordered pyrolytic graphite) presents a symmetry of order 3. The presence of lateral alkyl chains and their interactions with the graphite surface plays a major role in this kind of growing of fibrils.

7.4.2   Nanolithography

To replace the common optical photolithography (Figure 7.6) in the manufacturing of integrated circuits, new technologies are under development; these technologies

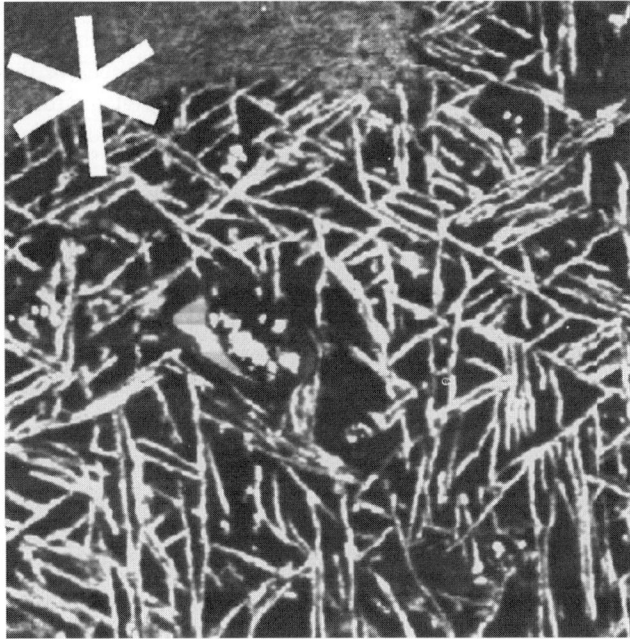

Figure 7.5   Atomic-force microscope picture of a deposit of polyfluorene on a layer of highly ordered pyrolytic graphite; the nanoribbons spontaneously turn in three directions

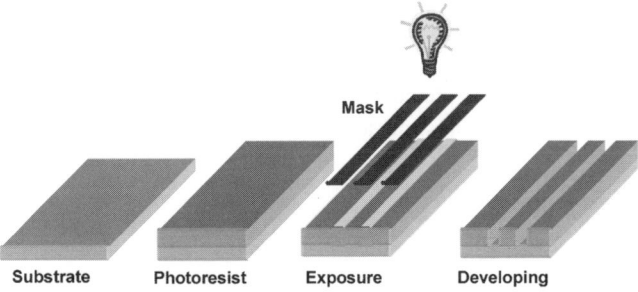

Figure 7.6   Principle of optical photolithography

intend to use rays of light corresponding to extreme ultraviolet (EUV) or X-rays. A particle beam, composed of electrons, can also be used to expose a photoresist sensitive to this light. This technology, known as *electron beam lithography* (EBL), has a variant: *electron beam projection lithography* (EBPL). Each of these technologies presents both advantages and drawbacks.

Lithography by X-rays is based on the use of a very short-wavelength radiation (smaller than a nanometre) and allows one to achieve the ultimate anticipated size of transistors on silicon, around 25 nm. In this case, an X-ray beam goes through a mask and directly exposes the substrate without going through a lens. Despite its intrinsic difficulties, the X-ray-based lithography is less expensive than the EUV-based technologies for the manufacturing of complex integrated circuits.

The extreme ultraviolet-based lithography is a technique under development that is favoured by manufacturers and is economically achievable in the coming years. Manufacturers anticipate that the EUV-based lithography will allow the fabrication of integrated circuits with transistors as small as 35 nm. However, this technique faces important difficulties because the wavelengths involved (1 to 40 nm) are refracted by any known material, in such a way that conventional lenses cannot be used.

It is also important to note that lithography is not used only to manufacture micro-processors, but also many other components, structures, and integrated circuits used in applications such as telecommunications, any kind of sensors, automobile industry, household appliances, or hi-fi.

7.4.3   Micro-contact printing

Micro-contact printing is a technique similar to those of common reprographics. From a structure made by conventional or EUV lithography, a mould known as a *stamp* is created. An elastomer material (often siloxane polydimethyle, PDMS) is deposited; after cross-linkage, the elastomer forms a matrix, which can be used to reproduce many examples of the initial nanostructure (Figure 7.7).

The motif reproduction can be achieved by stamping a self-assembled monolayer (SAM) and squeezing the mould on the layer deposited beforehand. Generally, small molecules (such as alkane thiols) are used; they are fixed beforehand on a thin layer of gold deposited on the substrate. The advantage of this technique is essentially that a large surface area (sometimes curved) can be rapidly structured, as illustrated in Figure 7.8. The self-assembled layer can then be used to protect the substrate (to be structured) from any chemical etching (Figure 7.9); one or many additional polymer layers are deposited on the substrate afterwards.

7.4.4   Ink-jet printing

This principle consists of forming droplets of uniform size from a fluid, which are pushed through a tiny nozzle. This technique has interested researchers for a long

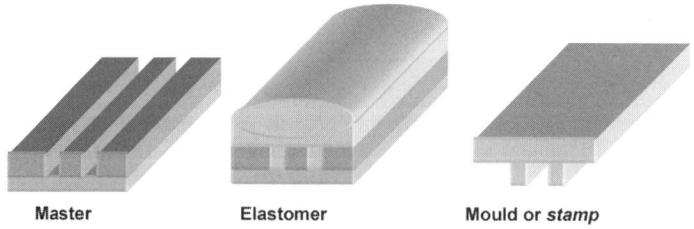

Master   Elastomer   Mould or *stamp*

Figure 7.7   Fabrication of an elastomer-based mould

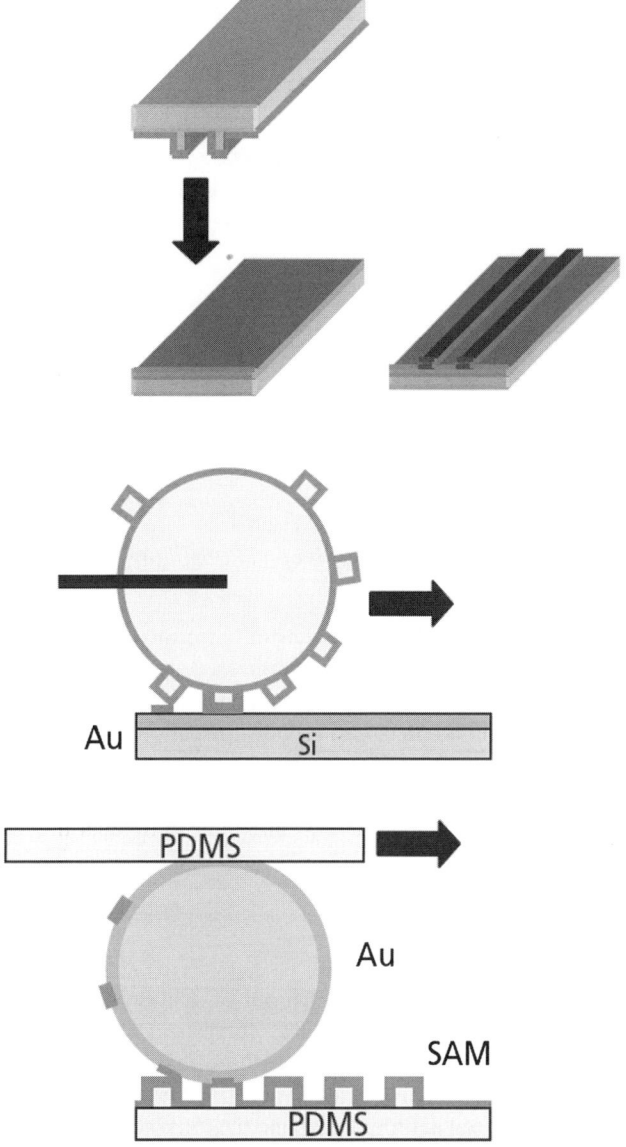

Figure 7.8   Reprographics principle of a nanostructured motif on a plane or curved substrate

time. The first works (from Felix Savart) date back to 1833 and the first mathematical description of the phenomenon was made by Lord Rayleigh. In that kind of system, a fluid under pressure can escape from a reservoir only through a hole of a few micrometres and a jet is transformed – under the effects of amplification of the waves

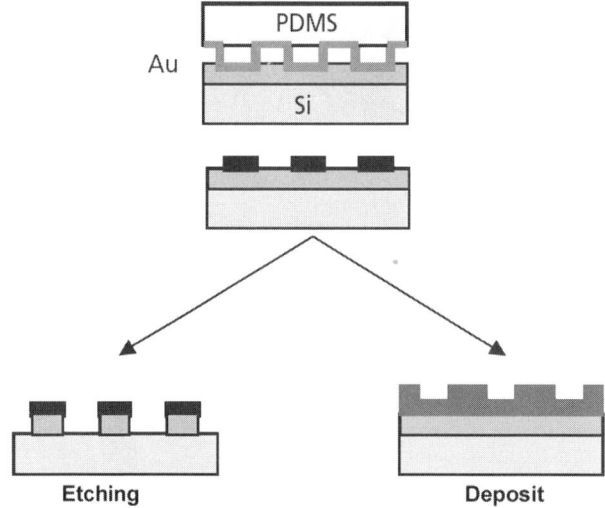

Figure 7.9   Chemical etching and deposit of additional layer(s)

of induced capillarity in the jet – into droplets of uniform size and shape. Generally, it is a piezoelectric device that generates the pressure waves that propagate in the fluid (Figure 7.10). The droplets split off from the jet thanks to an electrostatic field whose role is to charge the droplets during their formation. Once charged, they are directed towards a chosen spot on the substrate by another electrostatic field known as the *deflection field*. This kind of system is generally known as 'continuous system', since the droplets are produce continuously and their individual trajectory changes according to the applied charge.

7.5   Patterning

The behaviour and the shape of a liquid (in our case, a polymer in solution) on a surface depend on the long-range Van der Waals forces and the spreading (or wetting) coefficient S. In the process of wetting, there are three phases: solid, liquid and vapour (Figure 7.11). Three interfacial tensions are then defined: $\gamma_{LV} = \gamma$, γ_{SL} and γ_{SV}. They represent the energy required in order to expand the interface between two phases of a surface unit.

The wetting coefficient S is expressed hereafter:

$$S = \gamma_{SV} - \gamma_{SL} - \gamma \tag{7.1}$$

At the triple line, the liquid–vapour interface makes an angle θ with the solid plane. The smaller this angle, the easier the spreading. If the gravity effect is assumed to be negligible, the fluid drop equilibrium leads to Young–Dupre's law:

$$\gamma \cdot \cos(\theta) = \gamma_{SV} - \gamma_{SL} \tag{7.2}$$

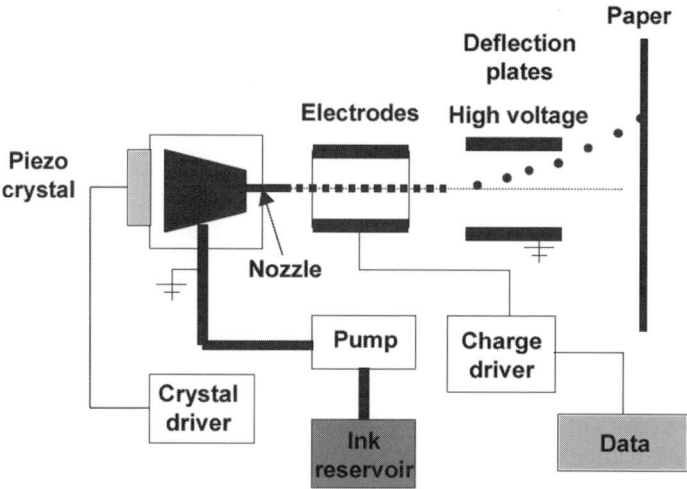

Figure 7.10 *Working principle of ink-jet printer head. It is exactly this principle that is used in commonly used ink-jet printers. This technique produces continuously droplets whose diameter is around twice the size of the nozzle at a frequency of 80 to 100 kHz. The development of tiny nozzles (which will allow nanometric droplets) is an essential factor for the fabrication of nanostructures using this approach, which at first sight looks easy to implement. The viscosity of the solution and the evaporation speed of the solvent are also parameters to master*

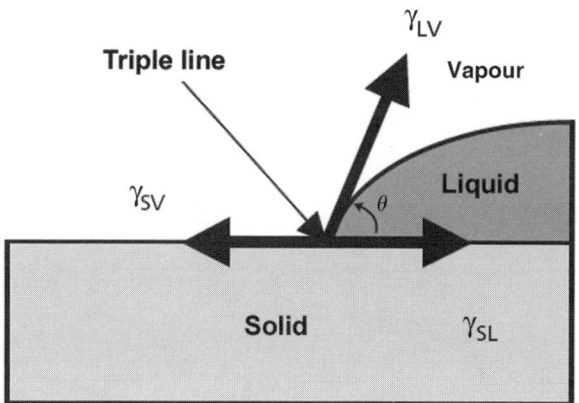

Figure 7.11 *Illustration of droplet angle*

Knowing the material used, there are essentially two theories that tell if the wetting is total, i.e. a fluid film is formed along the surface, or partial. One of these theories is based on the dielectric polarizability of the different phases in order to calculate the surface tensions and the spreading coefficient S. This approach allows one to determine how the wetability of a surface will be modified by the deposit of a layer. The second theory (based on Zisman's theory) is relative to the critical tensions. For a same substrate, the wetting angle θ is measured for liquids of a same group (*n*-alkanes for instance) and the different polarities. A straight line $\gamma = f(\theta)$ is obtained, whose extrapolation at $\theta = 0$ gives the critical surface tension γ_c. Consequently, fluids whose γ_{LV} is inferior to the γ_c of the solid lead to a total wetting.

'Dry' wetting is distinguished from 'moist' wetting roughly according to the volatility of the fluid involved. Dry wetting concerns fluids that are not very volatile. In that case, the final thickness of the film is ruled by the ratio S/γ. If this ratio is greater than 1, the order of magnitude of the thickness is one of a monomolecular layer; the liquid is completely spread. If the ratio is below 1, the thickness is macroscopic and its thickness can be measured via common optical methods. In the case of a moist wetting, the liquids have a high saturated vapour tension (it is the case for highly volatile liquids). There is always a film between the solid and the vapour phase, except far from the source of the liquid. However, the whole film is thermodynamically instable, given that its surface energy is not minimal and there is a critical threshold S_c. The different forces in competition create a fragmentation of the film when the thickness is smaller than a given limit. This is the aspect that is applied in the case being analysed: the formation of nanostructures. Under the effect of 'unwetting', the film that contains the polymer fractures and contracts and forms a typical nanostructure, more or less like a film of water drying on a pane of glass (Figure 7.12). A systematic analysis of the influence of the solution's viscosity, the film thickness, temperature and applied external constraints is essential to control the final size of the fabricated nanostructures (in a way that it is reproducible).

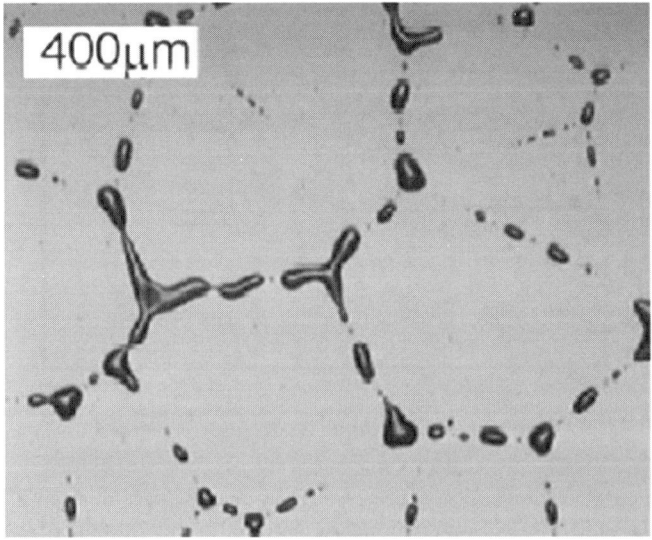

Figure 7.12   Unwetting of a water film on a pane of glass

7.6 Hybrid techniques

If the unwetting and stamping techniques are mixed, it leads to hybrid techniques such as micromoulding in capillaries (MIMIC) illustrated in Figure 7.13 and solvent-assisted micromoulding (SAMIN) illustrated in Figure 7.14. These techniques are increasingly used to produce nanostructures on plane or curved surfaces. The use of an appropriate solvent, which under capillarity forces fills up the mould deposited on the substrate beforehand, allows for the creation of nanodevices in polymer, e.g. polyacrylate or polyurethane. After filling up the mould with the solution, the polymer is reticulated, and then the mould is removed. In the SAMIN technique, the wetting of the mould surface is favoured to create the structures. Many examples of structures have been obtained via these two techniques, and the structures created in this way are used particularly in microfluidic devices.

7.7 Writing via local probe microscopy

An ideal way to produce nanostructures is to use the local probe microscopy technique by locally modifying the surface:

- by mechanically moving nanometric objects (nanoscanning of the surface);
- by removing or transforming molecules using an electric field (as when an STM tip that generates an electric field transforms the S_i-H bonds into S_iO_x), illustrated in Figure 7.15, or by a mechanical action that stamps a polymer surface (Figure 7.16), or even by optical methods that expose a photoresist film through the stretched optical fibre of a scanning, near-field, optical microscope (SNOM) as illustrated in Figure 7.17.

However, it is worth mentioning that these techniques – where local-probe microscopy is involved – suffer from a crucial lack of speed for the development

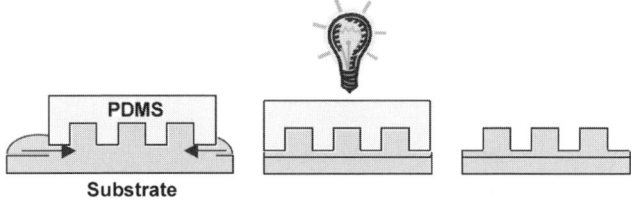

Figure 7.13   Micromoulding by capillarity (MIMIC)

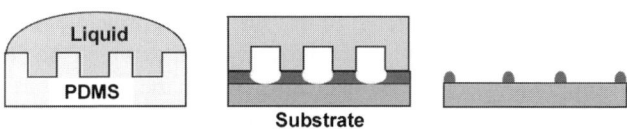

Figure 7.14   Micromoulding by the SAMIN technique

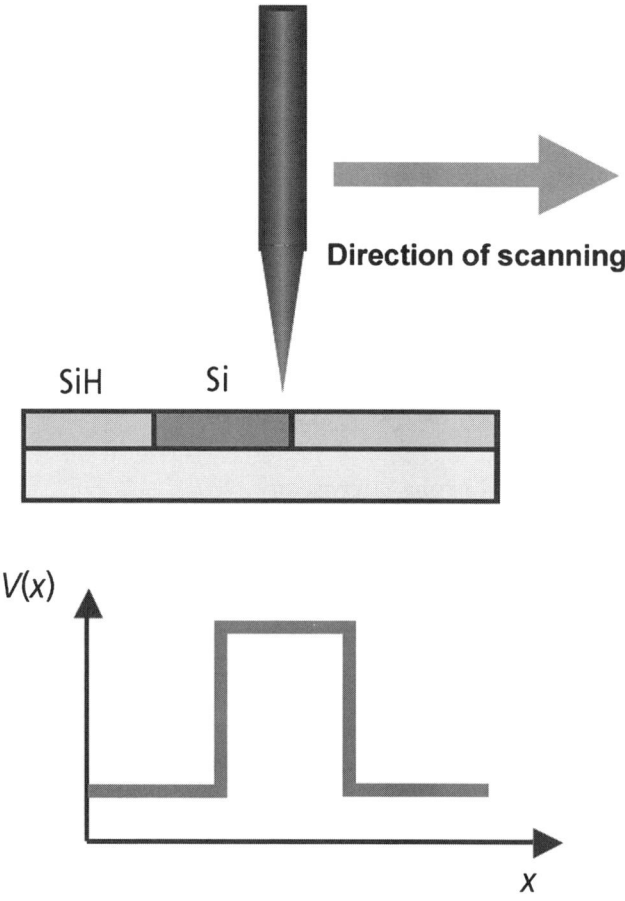

Figure 7.15 *Writing with a scanning tunnelling microscope (STM): the local appli-*
cation of an electric potential changes the chemical nature of the
layer

of nanostructures. The surface of the substrate is of course modified locally but the
modification is sequential. A parallelization of this process is under development,
particularly at IBM for the development of Millipede, a non-volatile computer memory based on a MEMS probe (Figure 7.18). It is a device composed of hundreds of
independent AFM tips, which can each modify the surface of the polymer material
or simply scan the topographical relief of this surface. Its principle is based on heat,
or, more precisely, on temperature threshold, according to which the device is in its
reading or writing mode. In the case of writing, where the temperature is high, this
allows it to indent (or print) the polymer film, since in this condition the material is
soft. In the case of reading, the temperature is decreased and the material hardens,
allowing the tips to read the information contained on the film without altering its

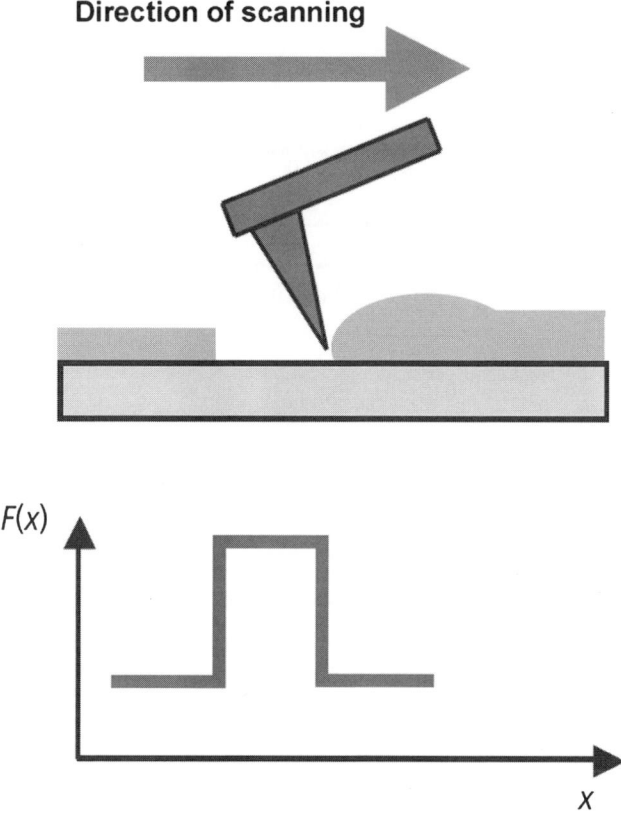

Figure 7.16   Writing with an atomic force microscope (AFM): the action of the tip locally removes matter where the applied force is greater

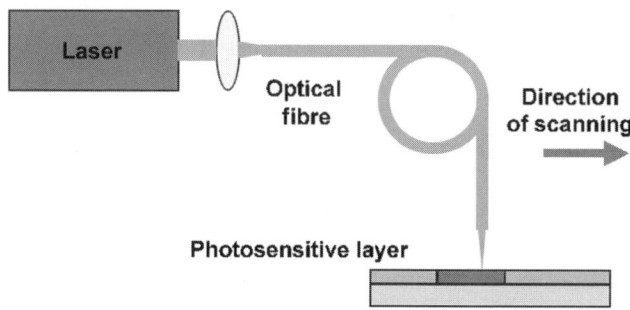

Figure 7.17   Writing with a scanning, near-field, optical microscope (SNOM): the action of light locally changes the photosensitive layer (resin)

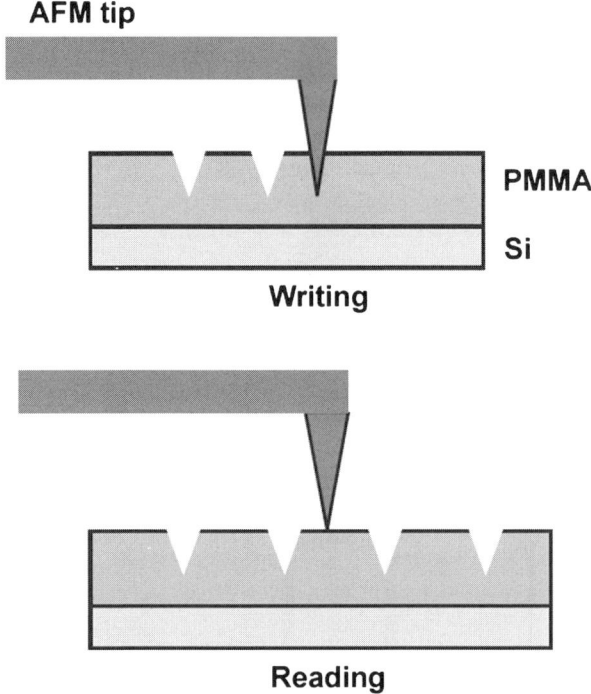

Figure 7.18   Working principle of IBM Millipede

content. In this case, the storage density is 20 times higher than the one obtained in the best current hard drive. IBM developed a prototype of the device, which it claims can hold the equivalent of 25 DVDs on an area the size of a postage stamp.

Another advantage of this technology is the low cost of fabrication, since it can be manufactured using existing processes. IBM demonstrated a prototype Millipede storage device at CeBIT 2005; this could be one of the different alternatives to the common hard drive. The system proposed by IBM allows, theoretically, a much larger storage capacity at a lower price. Another advantage compared with a hard drive is a greater resistance to shocks or to heat, which are essential parameters in mobile devices.

7.8   Design and development of molecular circuits

Therefore, as long-term logical consequences and perspectives of these works, scientists try to develop systems for electronic data processing at the molecular level. The development of these systems will contain the following stages:

- the design of electrical molecular wire capable of carrying data; these wires must be made of conjugated systems long enough to cross a typical molecular element such as a mono- or bi-layer membrane;

- the possibility of attaching these wires to a membrane or a surface in order to connect this molecular devices; it is here a matter of making a macro-contact with the device in order that the device can be controlled via simple commands (electrical potential, irradiation);
- the attachment to these wires of molecular systems capable of sensing, processing or storing the data generated by an external stimulus and transmitted by these wires; here it is a matter of building a set of interconnection nanocontacts between the molecular wires and the functional molecular entities and inducing a physicochemical response to this stimulus;
- the design of the whole system as a parallel arrangement of molecular electric devices whose performances would therefore be independent of the possible failures of one of the devices;
- finally, the organization of these wires into 'circuits' or 'rows' by linking to a device or by self-assembling based on molecular recognition phenomena.

It is worth noting that, by bringing together physicists, electronics specialists and chemists via their specific skills and knowledge, this theme (which has been developed until now separately by each community) would allow, in the near future, the composition of a single high-performance device. Without any interdisciplinarity, this promising field of research could not succeed.

Chapter 8
Organic-matrix-based nanocomposites

For a long time, humankind has tried to improve the properties of the materials of its environment or those that it has fabricated. From very early, it has mixed different substances to improve the properties of the resulting material. The results of this process are called *composite materials.*

Composite materials include all close combinations of immiscible materials (heterogeneous structures) that give to the resulting material properties that none of the initial materials individually possessed.

Under this definition a large range of products can be found. It includes archaic material such as cob – a house-building material consisting of clay, sand, vegetable fibre (e.g. straw) or animal fibre (e.g. wool or hair), water and earth, which was used during Neolithic times; and some of the most modern materials, which combine carbon or glass fibre with synthetic resins for the fabrication of a great number of devices requiring mechanical resistance and lightness.

Nature has also used this intimate combining of various substances within a single material such as wood (cellulose fibres reinforcing a lignin-based matrix) or bone. Bone is a composite formed by a calcium phosphate (hydroxyapatite) and protein (collagen). The collagen fibres compose a framework on which the hydroxyapatite crystals gather. These fibres hold the hydroxyapatite crystals together, set the orientation of the atomic stack in the crystals, and consequently increase the mechanical resistance of the bone.

In the light of these examples, in most cases, a composite can then be described as a dispersion (organized or not) of one or many particle materials or fibre materials – known as *dispersed phase* or *reinforcement* when the intended property is mechanical resistance – in a material used as a matrix binder.

A nanocomposite is a composite material whose dispersed phase constitutes particles at least one of whose free dimensions is in the order of a nanometre or a few dozen nanometres at maximum. Under this definition, many metal-matrix-based nanocomposites are gathered (for instance materials resulting from the dispersion of nanoparticles of oxide, nitride, or metal carbide in matrices or metal alloy), ceramic or inorganic matrices-based nanocomposites, or nanocomposites based on organic matrices. The last group is described in this chapter, with a particular focus on polymer matrices[1].

8.1   Types of nanoparticle

Nanocomposites can be classified according to the morphology of nanoparticles used and particularly according to the number of their nanometric dimensions.

8.1.1   Nanoparticles with three nanometric dimensions

In this group, there are a great number of nanoparticles whose properties have been described in other chapters, such as the *atom aggregates*, the *amorphous* or *crystalline metal nanoparticles* (gold, platinum, silver, copper, iron, etc.), the fullerenes (spherical or pseudo-spherical allotropic variety of carbon), the *isometric nanoparticles* derived from sulphide, selenide, nitride or carbide oxide for instance, particles of magnetite (Fe_3O_4), of cadmium sulphide (CdS) or cadmium selenide (CdSe). A mechanical reinforcement effect is not usually desired, but instead particular attributes such as optical, conducting and magnetic properties, when these nanoparticles are distributed within an organic material whose shape can easily be controlled.

8.1.2   Nanoparticles with two nanometric dimensions

When the two dimensions of the nanoparticle are in the order of a few nanometres, the third one being much greater, the devices are hollow (nanotubes) or solid (nanowires, nanowhiskers). Among the nanotubes, the best known are the single- or multi-wall carbon nanotubes (see Chapter 2). Boron nitride nanotubes also exist. They are isoelectronic structures of carbon nanotubes and natural nanotubes, in the form of tubular aluminosilicate (imogolite).

Regarding nanowhiskers, a lot of work has been done on these kinds of materials, applying the concept to a wide range of chemical compounds (metals, oxides, arsenide, silica); but few developments have been made in their function of dispersing charge in polymer matrices. Only cellulose nanowhiskers (an extract from the carapaces of certain shellfish) and phlogopite nanowhiskers (a solid tubular aluminosilicate) have been tested as reinforcement in polymer matrices.

8.1.3   Nanoparticles with a single nanometric dimension

When the nanoparticle has only one single nanometric dimension, the other two dimensions being more than 100 nm, the geometry of this particle is of the type of nanolayer. This nanoparticle's group comes from materials that are naturally laminated, whose principal groups are presented in Table 8.1. Most of these nanolayers

---

[1] Polymer: a substance composed of large molecules formed by the repetition (or sequence) of the same motif composed of one ore many base units, also known as *monomers*. The mean number of these monomers in the final macromolecules represents the polymerization degree. If this polymerization degree is large, the polymer is known as *'high' polymer*; when the degree is low, the polymer is known as an *oligomer*. Homopolymers are composed of one single type of monomer, which motif is repeated in the molecule; copolymers are composed at least of two types of monomer.

Table 8.1 Principal groups of materials preceding the nanolayers

Precursors of the nanolayers	Examples	Electrical charge of the layer
Elements	Graphite	neutral
Halides and metal cyanides	MX_2(M = Cd, Ni, Pb; X = Cl, CN, I)	neutral
Metal chalcogenides	MX_2 (M = Mo, Sn, Ta, Ti, V, W, Zr; X = S, Se)	negative
	MPX_3 (M = Cd, Co, Fe, Mg, Mn, Zn; X = S, Se)	
Carbon oxides	Oxided graphite	negative
Metal oxides	MoO_3, V_2O_5, $MOXO_4$ (M = V, Nb, Ta; X = P, As)	negative
Metal phosphates	$M(HPO_4)_2$ (M = Ce, Sn, Ti, Zr, Hf)	negative
Niobates and titanates	$KNbO_3$, $K_4Nb_6O_{17}$, $K_2Ti_4O_9$, $H_2Ti_3O_7$, $KTiNbO_5$	negative
Smectites and silicate layers	Montmorillonite, (fluoro)hectorite, saponite, magadiite	negative
Double-layer hydroxides	$M_xM'_y(OH)_a(anions)_b \cdot cH_2O$ (M = Fe^{3+}, Al^{3+}, Cr^{3+}; M' = Mg^{2+}, Co^{2+}, Fe^{2+}, Ni^{2+}, Cu^{2+}); $LiAl_2(OH)_6 \cdot 2H_2O$	positive

are characterized by the presence of charges (positive or negative) localized at the surface.

The most common nanoparticles used for the preparation of matrix polymer-based nanocomposites are, on the one hand, double-layer hydroxides and, on the other hand, swelling clays also known as smectites.

Double-layer hydroxides are composed of a mix of $M_xM'_y(OH)_a(anions)_b \cdot cH_2O$-type layers, characterized by layers positively charged, due to a defect of hydroxide counterions, which could be localized in the crystalline structure of the layers, and counterbalanced by anions (hydroxyls, carbonates, chlorides…) localized in the space between two layers.

Smectites are principally represented by montmorillonite, an aluminosilicate whose backbone is based on an alumina octahedron sandwiched between two silica tetrahedrons. During the geological formation of this montmorillonite, some of the backbone atoms are replaced by atoms of the same size but with lower charge. Therefore, the silicon (4^+) can be replaced by iron (3^+) or aluminium (2^+), and the aluminium (3^+) of the backbone can be replaced by iron (2^+) or magnesium (2^+).

Therefore, these layers, stacked up, are globally negative and these negative charges are counterbalanced by cations (Na+ or Ca++, for instance), highly hydrated and localized in the spaces between layers, known as *galleries* or *interlayer spaces*. The montmorillonite's structure is illustrated in Figure 8.1.

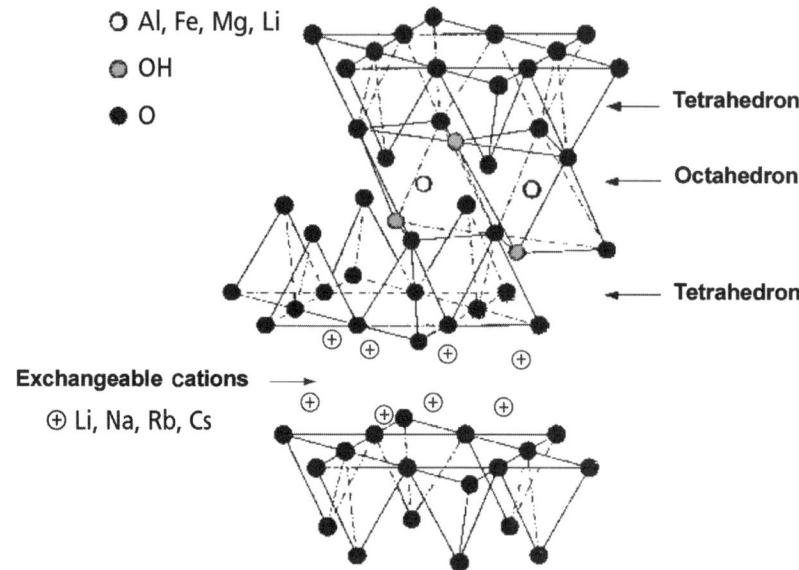

O Al, Fe, Mg, Li

◎ OH

● O

Tetrahedron

Octahedron

Exchangeable cations →

⊕ Li, Na, Rb, Cs

Tetrahedron

Figure 8.1   Structure of montmorillonite

Nanocomposites based on smectites such as montmorillonite aroused a great interest in the academic and industrial worlds, due to the importance of modifying properties obtained in the organic matrices even with low smectite concentrations. They currently represent the only commercial development in the field of organic matrix-based nanocomposites.

8.2   Preparation of nanocomposites

In order to scatter nanoparticles in an organic medium, the first difficulty to over-come is avoiding or destroying particle aggregations within the medium. Whatever the morphology of nanoparticles, they generally tend to form aggregates stabilized by various types of interaction (ionic interactions, hydrogen bridges, Van de Waals forces, etc.) that are often more important than interactions between the surface of the particles and the organic matrix in which they are scattered.

For instance, layers of montmorillonite are stabilized by the attraction of globally anionic layers with cations located in the interlayer space. Moreover, these cations, which are highly hydrated, lead to the formation of a highly hydrophilic space between the layers, preventing the penetration of a great number of organic molecules, which are often hydrophobic. Consequently, it is essential – in order to make the interlayer space of these montmorillonites more hydrophobic and to reduce the Coulombian interactions that hold the structures stacked – to exchange these highly hydrated inorganic cations with more hydrophobic organic cations. Classically, ammonium

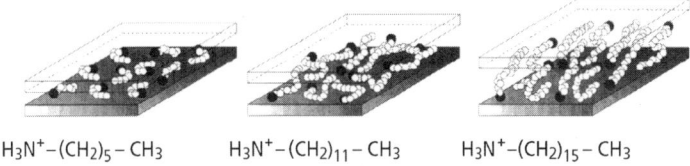

$H_3N^+-(CH_2)_5-CH_3$ $H_3N^+-(CH_2)_{11}-CH_3$ $H_3N^+-(CH_2)_{15}-CH_3$

Figure 8.2   Illustration of montmorillonites organically modified by ammonium cations with alkyl chains of increasing length

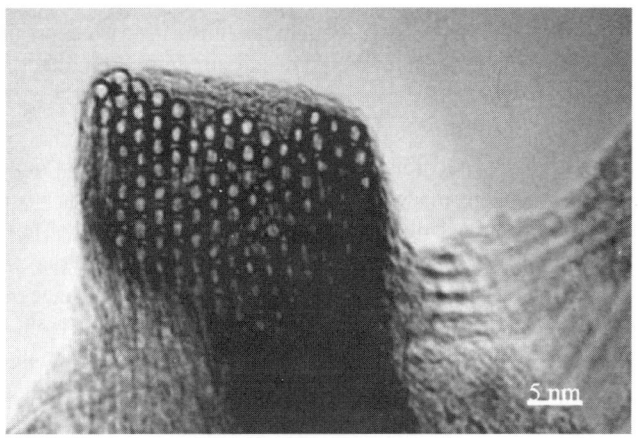

Figure 8.3   Transmission electronic microscope image of a transverse section of a rope of mono-walled nanotubes [Source: Thess et al., Science, 273:483–7, 1996 American Association for the Advancement of Science]

cations with long alkyl chains are used; the presence of long alkyl chains ensures the hydrophobization of the interlayer space and noticeably reduces the attractive Coulombian interactions between two adjacent layers by increasing the occupied volume.

This exchange is easily done in the aqueous phase, where the presence of water solvates enough inorganic cations initially present in the interlayer spaces to swell the clay to its maximum size.

In the case of carbon nanotubes, Van der Waals interactions, such as those observed between graphite layers, cause the aggregation of nanotubes into ropes (Figure 8.3).

Here again, the use of compatibilizing agents and indeed the chemical grafting of organic molecules turns out to be essential for adequate dispersion in an organic matrix such as a polymer.

In the case of cellulose nanowhiskers, the essential interactions to break are the interactions of hydrogen bridges between nanoparticles. In this case, the main technique consists in dispersing the cellulose nanowhiskers in water, where the inter-particle hydrogen bridges vanish, and putting this suspension in contact with a latex polymer. The composite is then formed, either as a film after evaporation (of the water) or as a powder after lyophilization, which will be dispersed in molten materials (see below).

Finally, the technique most widely used to prevent aggregation, in the case of isometric nanoparticles, is stabilization by use of surfactive or reactive molecules, i.e. those with a hydrophilic or reactive 'head' able to interact with the hydrophilic or polar surface of the nanoparticle, and a long hydrophobic 'tail', allowing the dispersion in a relatively hydrophobic matrix.

8.2.1   *Dispersion of nanoparticles in a preformed polymer matrix*

From a preformed polymer matrix, two techniques can be used to ensure the dispersion of nanoparticles: dispersion in solution and dispersion in molten materials.

8.2.1.1   Dispersion in solution

The first attempts for the preparation of nanocomposites based on montmorillonite and a polymer matrix were made by dispersing clay in a polymer solution, in an appropriate solvent, followed by the evaporation of the solvent. The results were often disappointing. In most cases, the polymer does not penetrate the interlayer space, or does so only partially. This failure is ascribed to the presence of the solvent, as it is this that competes with the polymer. The polymer in solution is characterized by many configurations of chains, which are free to move around. When this is confined in the interlayer space, it loses its conformational freedom, which is expressed by a heavy loss of entropy, and is unfavourable to the penetration of the polymer. The solvent, being a small molecule, is not subjected to this entropy loss and penetrates the interlayer space preferentially.

In the other cases (nanotubes and isometric nanoparticles), this procedure does not lead to any entropy modification and thus is suitable for the preparation of nanocomposites, especially when the desegregation is improved by the use of ultrasound. This method has been used for the preparation of nanocomposites based on carbon nanotubes dispersed in matrices of polystyrene or polyvinyl alcohol.

8.2.1.2   Dispersion in molten materials

The thermoplastic polymer matrices have the property of behaving as a viscoelastic fluid, beyond a critical temperature (the temperature of vitreous transition for amorphous matrices, and the temperature of fusion for semi-crystalline polymers). In this state the polymer can be dispersed, i.e. mixed using appropriate tools (mixer, extruder, etc.). In 1993 work conducted by Cornell University (New York) demonstrated the potential for 'molten' polystyrene to penetrate within the interlayer space of a montmorillonite that has been modified by a quaternary ammonium. Since then, this technique has been successfully used for other polymer matrices such as poly(methyl-methacrylate), poly(ethylene-oxide), vinyl ethylene-acetate copolymers, etc.

Some polyolefines that are not originally sufficiently polar (e.g. polyethylene or polypropylene) cannot penetrate the interlayer space of a montmorillonite when modified by an ammonium with long alkyl chains. However, the addition of a small quantity of maleic-anhydride-grafted polyethylene (MAHgPE) or maleic-anhydride-grafted polypropylene (MAHgPP) – with a level of maleic anhydride low enough to prevent making this additive immiscible with the polyolefine – allows the formation of a nanocomposite through penetration of the polymer mixture in the interlayer space.

This technique of dispersion has been successfully used in the case of multi-walled carbon nanotubes in polypropylene matrices, also in the case of vinyl ethylene-acetate with cellulose nanowhiskers in vinyl polychloride matrix.

8.2.2 Matrix synthesis in the presence of nanoparticles

Another technique allowing the preparation of nanocomposites consists of the polymerization of monomers in the presence of nanoparticles. This kind of preparation in particular allows the creation of nanocomposites based on thermally stiff matrices such as epoxy resin and polyurethane resin. Additionally, it facilitates grafting between the nanoparticle and the polymer matrix, in the case where the nanoparticle has been modified either by a chemical reaction capable of polymerizing, or of starting or catalyzing the polymerization.

The fixation of a Ziegler–Natta catalyzer (active in ethylene polymerization) at the surface of nanoparticles of Cr-Fe, has allowed the fabrication of nanocomposites of magnetic nanoparticles dispersed in a polyethylene matrix (Figure 8.4).

This technique has also been used for the preparation of nanocomposite based on carbon nanotubes, by polymerization of styrene and grafting of the polystyrene at the surface of the nanotubes by radical reaction.

Historically, it is also this technique that was used by researchers at Toyota in 1989 to create the first nanocomposites based on montmorillonite. Ammonium with carboxylic acid functions was used to replace the inorganic cations of natural montmorillonite. These organically modified montmorillonites were heated in the

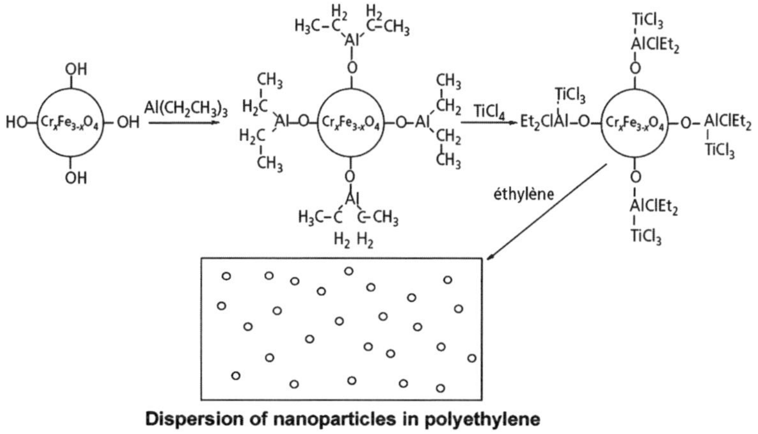

Dispersion of nanoparticles in polyethylene

Figure 8.4   Preparation of a nanocomposite based on magnetic nanoparticles dispersed in polyethylene

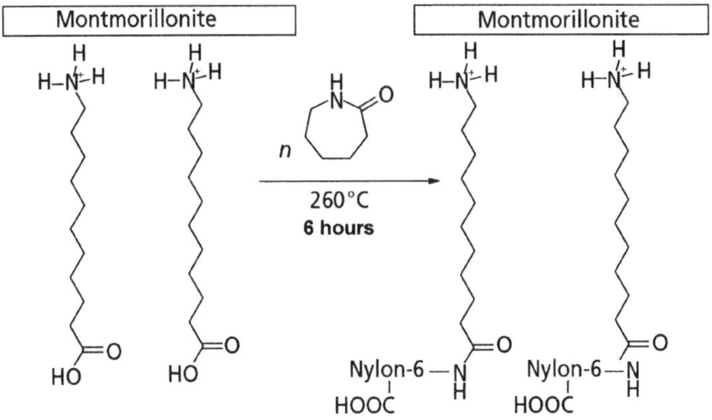

Figure 8.5   Preparation of montmorillonite/nylon-6 nanocomposites

presence of ε-caprolactame, leading to the polymerization of this monomer to form the corresponding polyamide, nylon-6 (Figure 8.5).

From that time forward, this polymerization technique in the presence of organophilic montmorillonite ˚has been successfully used for a great number of monomers, giving access to many nanocomposites based on a polymer matrix (such as polyacrylates, polyamides, polyanilines, polyesters, polyolefines, polystyrenes, etc.) and theoretically allowing the creation of nanocomposites with any type of polymer matrix. It is absolutely essential to adapt the choice of the modifying organic agent (i.e. the ammonium cation with long alkyl chains) to the chosen monomer and consequently to the *in situ* generated polymer matrix. The nanometric dispersion of nanolayers primarily imposes the insertion of the monomer within the clay gallery in the organically modified surface. Subsequently, it imposes their polymerization (thermally or by catalysis) leading to a delamination or even the exfoliation of silicate layers; this depends on the monomer concentration in the growing polymer matrix (see below).

8.2.3   *Preparation of nanoparticles in organic matrix*

This technique, which is relatively marginal for nanolayers and unused for nanotubes and nanowhiskers, has been used for the preparation and stabilization of isometric nanoparticles. For instance, the heating of a copper (II) formate and poly-2-vinylpyridine (complexing agent of copper) mixture, to 125 °C, leads to an oxido-reduction reaction between the Cu^{2+} ions and the formate ions, with the reduction of Cu^{2+} to metallic copper. Analysis of the films generated shows the formation and the dispersion of copper nanoparticles of 3.5 nm with a mass concentration of 23 per cent.

A similar technique has been used to allow dispersal of gold nanoparticles. Copolymers based on polystyrene-*b*-poly(2-vinyl pyridine), forming micelles full of pyridine units in non-polar solvents, are used to complex a gold salt such as $HAuCl_4$.

Caused by the heat and the effect of hydrazine N_2H_4, the gold salt is reduced to a gold nanoparticle, the size of which is determined by the size of the complexing micelle (around 10 nm), then stabilized and distributed in the polymer matrix.

8.3   Characterization and properties

8.3.1   *Morphological characterization: tools and techniques*

The principal technique for morphological characterization of organic-matrix-based nanocomposite is transmission electronic microscopy (see Appendix 1). This technique shows the morphology and the state of the nanoparticle dispersion in the organic matrix, from thin cuts of the nanocomposite. Classically, nanoparticles appear in dark on a light background, since the atoms that usually make up most nanoparticles are heavier, and therefore more easily halt the transmission of electrons compared with those of the hydrocarbon matrix.

In the case of nanocomposites based on nanolayers, besides the homogeneous dispersion of nanoparticles in the organic matrix, other morphologies can develop. During the penetration of the polymer in the interlayer space, a stable and regular structure – characterized by a stack alternating between nanolayers and a single layer of polymer – can be formed. This is known as *inserted nanocomposites* and *insertion* phenomenon (Figure 8.6a). In contrast, the situation where the homogeneous dispersion is obtained is known as *exfoliation* or *delamination* and leads to *exfoliated* or *delaminated* nanocomposites (Figure 8.6b).

These morphologies are easily shown by transmission electronic microscopy, but another technique appears to be interesting for the characterization of inserted nanocomposites: X-ray diffraction. This technique allows measurement of the distance between the surfaces of two adjacent nanolayers. While a natural montmorillonite layer may have a nanolayer gap of around 1.2 nm, its modification by an alkyl ammonium cation can lead to a significant increase in this gap, which varies from 1.5 to more than 3 nm, in relationship to the length of the alkyl chain. The

(a)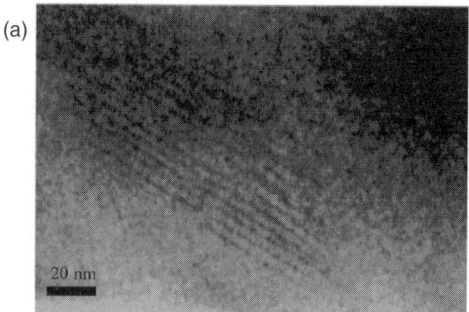

(b)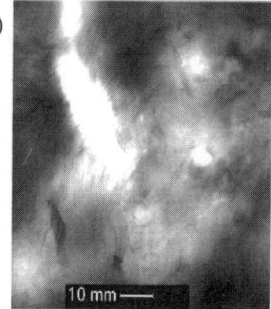

Figure 8.6   *Transmission electronic microscopy pictures of nanocomposites based on montmorillonite layers in a thermoplastic matrix: (a) inserted structures; (b) exfoliated structures*

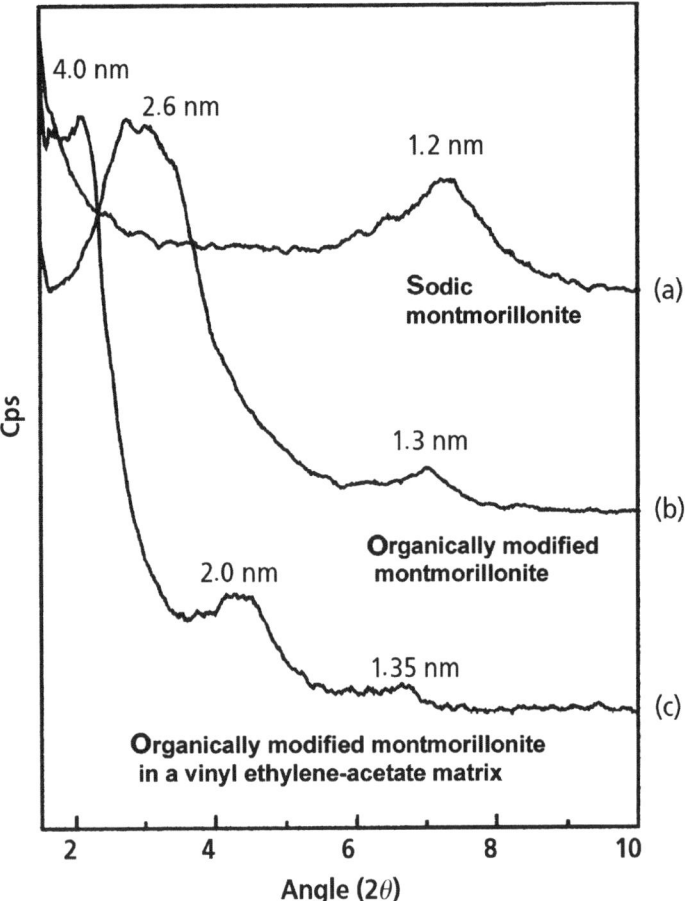

Figure 8.7   *Diffractogram X: (a) sodic montmorillonite; (b) after exchange with a dimethyldioctadecylammonium; (c) after insertion of a vinyl ethylene-acetate copolymer (3 per cent of montmorillonite)*

insertion of a monolayer of polymer leads to an additional increase of 0.4 to 1 nm, dependent on the structure of the particular polymer. This technique is illustrated in Figure 8.7, which shows the diffractogram X of a sodic montmorillonite, the same montmorillonite after modification by a dimethyldioctadecylammonium, and finally the nanocomposite obtained by redispersion of the organically modified smectite in a matrix of vinyl ethylene-acetate copolymer.

Note that the inserted structure and the exfoliated structures are two extremes and, in many cases, the real structure (such as shown by transmission electronic microscopy) is usually in an intermediate state where an insertion can be seen by X-ray diffraction, as can the presence of exfoliated layers or small clusters of layers composed of two to a dozen inserted nanoparticles. More rarely, other structures, such as a regular stack similar to the insertion, but with distance between layers greater than 7 nm (the minimal detectable size with X-ray diffraction) can be seen.

8.3.2 *Properties*

The analysis of nanocomposites based on isometric nanoparticles has not yet fully developed. The expected properties are essentially linked to the *specific* properties resulting from the nanometric scale. Regarding nanotubes, nanowhiskers and nanolayers, many interesting properties result from their dispersion in an organic polymer matrix.

8.3.2.1 Mechanical resistance

Mono-walled carbon nanotubes have a high tensile strength. For illustrative purposes, let us say that a carbon nanotube of 1 nm diameter is six times more rigid (1 200 GPa), in the longitudinal direction of the nanoparticle, than steel (200 GPa). Its tenacity (maximum applied stress that would break the material) is of 200 GPa, much higher than the steel tenacity which is 1.5 GPa. Moreover, these carbon nanotubes are also very resistant to compression, flexion and torsion. This makes them very attractive for the reinforcement of polymer materials. Therefore, the insertion of nanotubes in a polystyrene matrix, in combination with the use of tensioactive agents to limit their self-aggregation, allows an increased rigidity in the nanocomposite of about 30 per cent with only 1 per cent addition of mono-walled nanotubes, which is amazing considering such a small quantity of reinforcement. However, studies show that the tensile strength of a mono-walled nanotube is less astounding. With this kind of material, the nanotubes tend to slide along each other within the nanoparticle itself, consequently reducing the reinforcement effect that is suggested by the intrinsic rigidity of the nanotubes when considered individually.

Cellulose nanowhiskers also show good properties of reinforcement with an abrupt increase in the rigidity beyond a certain weight ratio (between 1 and 5 per cent, depending on the polymer matrix under consideration). This abrupt increase in a property comes from a phenomenon known as percolation, i.e. the formation of a three-dimensional lattice by interaction between the nanowhiskers in mutual contact. The rigidity of this lattice is ensured by both the intrinsic rigidity of nanowhiskers (150 GPa) and the formation of hydrogen bridges during contact with the next nanowhisker. The rigidity of the material – beyond its vitreous transition (i.e. in the non-vitreous state of the polymer) – can be increased up to 1 000 times, during the crossing of the percolation threshold.

When considering the nanolayers, a reinforcement in the rigidity of the nanocomposite material is observed in case of exfoliation. Generally, the rigidity of a sample subject to a tensile load increases by a factor of 2 for a weight ratio of around 3 per cent of inorganic nanolayers. Beyond this insertion ratio, the increase in rigidity as a function of the quantity of nanolayers is less important. This is easily explained by purely geometrical constraints. Let us consider the homogeneous dispersion of these nanolayers in a given volume. The homogeneous distribution in the volume is possible only for a limited number of layers (this being a function of the size of the nanolayers).

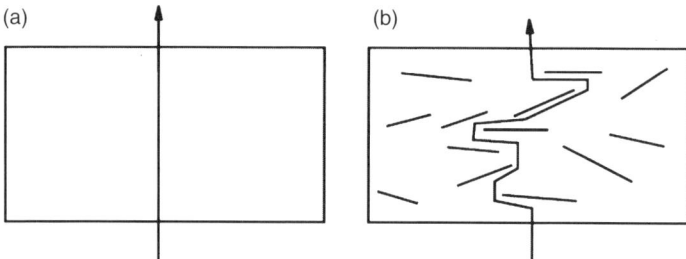

*Figure 8.8   Barrier effect: illustration of the barrier effect in an exfoliated nanolay-
ered nanocomposite: (a) matrix alone; (b) nanocomposites based on
exfoliated nanolayers*

Beyond this limit, the presence of a layer next to another layer will automatically lead
to a reorientation of the latter through lack of space. In the case of a smectite-type
nanocomposite based on clay, this is expressed by a tendency to limit the exfoliation
in favour of an inserted structure or at least an organized one, when the insertion ratio
exceeds the critical value.

8.3.2.2   Fluid barrier properties

Nanocomposites based on exfoliated nanolayers, because of their structures, have
another interesting property, which consists of an important reduction of fluid diffusion (gases, liquids) throughout the material: the 'barrier' effect. The origin of this
property is illustrated in Figure 8.8.

In the example (a), the molecule diffuses directly throughout the homogeneous
matrix and its speed is only a function of the diffusion coefficient of this molecule
in the polymer matrix. In the example (b), the exfoliated nanolayers, which are
completely impermeable to molecules diffusing in the material, behave like barriers
that these molecules must skirt around in order to progress on their way. The distance
to cover going through the material is significantly increased, which leads to an
important kinetic reduction in the diffusion of molecules within the material. For a
nanocomposite with a weight insertion of 3 per cent of exfoliated nanolayers, the
diffusion throughout the film can be reduced by a factor of 5, compared with the
same pure polymer film.

8.3.2.3   Fire resistance properties

Nanocomposites based on exfoliated nanolayers also allow an improvement in the
fire resistance of polymer materials. To understand this characteristic, it is essential to
illustrate briefly the combustion mechanism of an organic material. When an organic
material is heated beyond a certain temperature, some chemical bonds of the material
break, leading to the formation of volatile materials (the *fuel*), which emanate from
the material. Furthermore, this degradation is favoured if oxygen can penetrate the
heated material (this phenomenon is known as *thermo-oxidation*). In the presence

of atmospheric oxygen, this fuel can spontaneously ignite, in direct contact with the surface of the material. The energy produced by this combustion – which is only a series of chemical reactions involving highly reactive radicals – helps to maintain and increase the heating of the material, until the consumption of fuel is complete.

The fire resistance, as it is observed in exfoliated nanolayered based nanocomposites, is due to two phenomena. The first one results directly from the barrier-to-fluid properties described above. The presence of nanolayers significantly reduces the diffusion of the fuel towards the outside, and the diffusion of oxygen towards the inside, consequently limiting the degradation and the heating of the material. The second phenomenon results from the reorganization of the layers during the degradation of the material. Under heat and thermo-oxidation degradation, the exfoliated nanolayer based structure collapses to form a crust of nanolayers stacked at the surface of the sample. This collapse leads to the trapping of material in decomposition, which gets caught between the layers and is incapable of undergoing an effective thermo-oxidation. A real protective layer is therefore created at the surface, forming a thermal insulator and reducing the quantity of heat resulting from the flame, essential to maintain the thermal degradation.

In a nanocomposite with a 3.5 per cent weight of nanolayer of exfoliated montmorillonite in a vinyl ethylene-acetate copolymer matrix, a shift in the degradation temperature of the polymer matrix of more that 50 °C (in ambient air) can be observed during a heat increase of 20 °C/min; a reduction of 45 per cent of the maximum heat released by the combustion of the material is also observed at the same time. Furthermore, the protecting layer (dark crust) generated at the surface of the material in combustion, resulting from the collapsing of layers of clay, prevents the material from leaking out with the formation of burning droplets capable of propagating the fire to surrounding materials. This behaviour is shown in Figure 8.9 with a reduction in the propagation of flame and the absence of burning droplets in the case of vinyl polymer based on ethylene-acetate copolymer with 3.5 per cent in weight of nanolayers of exfoliated montmorillonite.

Some studies have shown that carbon nanotubes, dispersed in a polypropylene matrix, are also capable of improving the fire resistance of polymer materials. The mechanism leading to this property, in the case of carbon nanotubes, is still not well understood.

8.3.2.4 Other properties

Other properties specific to certain types of nanoparticle can also be used. The nanolayers of graphite and some carbon nanotubes are excellent electrical and thermal conductors. Some preliminary works show the importance of these nanocomposite conductors for electricity.

Optical properties of isometric metallic nanoparticles or sulphide are also interesting. Thus, it is known that gold nanoparticles dispersed in a medium emit different colours depending on their aggregation state. The dispersion of gold nanoparticles in polyethylene films of ultra-high molecular weight, followed by the stretching of the film along a certain axis, allows the production of an anisotropy in the distribution

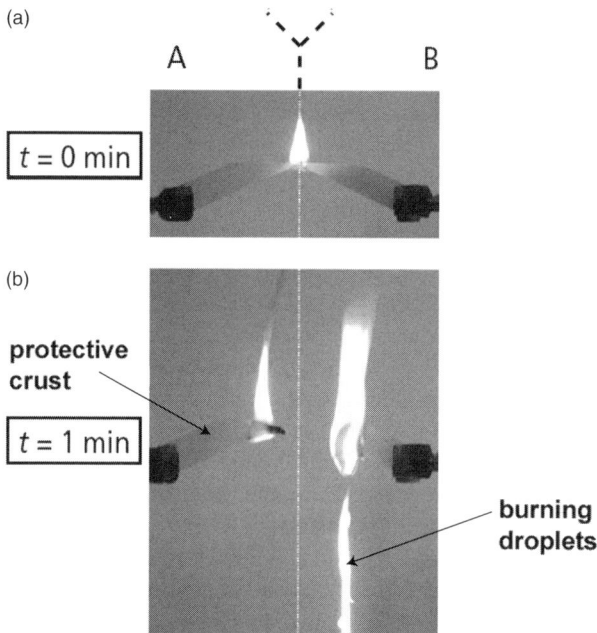

Figure 8.9   Resistance to fire of a vinyl ethylene-acetate based copolymer: (a) nanocomposites with an insertion of 3.5 per cent in weight of nanolayers of exfoliated montmorillonite; (b) matrix alone (t = 0 min, corresponds to the time where the material is lit)

of the particles within the film. Therefore, these particles preferentially aggregate perpendicular to the direction of stretch rather than parallel to it. According to the film orientation, regarding a source of polarized light, the film then shows a different colour, resulting from the variation in aggregation state of the nanoparticles.

Lead sulphide nanoparticles of nanometric size (between 2 and 40 nm) distributed in a polymer matrix (for instance ethylene polyoxide), in a proportion equivalent to 50 per cent in volume (i.e. 90 per cent in weight) produce a transparent material of very high refractive index (around 3), because lead sulphide nanoparticles are too small to diffract light whose wavelength is between 400 and 700 nm.

8.4   Applications

From the various nanocomposites presented previously, the materials based on exfoliated smectite nanolayers are already in use in industrial applications.

8.4.1   Nylon-6 smectite nanocomposites for clingfilm

Many industries produce clingfilms based on nylon-6 and exfoliated montmoril-lonite nanocomposites, as clingwraps for food with strong odour, such as cheese, and food sensitive to oxygen, such as meat. The excellent barrier properties of these nanocomposites are being utilized here. Weight concentrations in montmorillonite layers of 3 per cent lead to a reduction of permeability by a factor of 2, while at the same time keeping the same transparency of the film and increasing its stiffness by a factor of 2. These films are obtained by the method of *in situ* polymerization, as described in Figure 8.5.

8.4.2 Nanocomposites based on vinyl ethylene-acetate copolymer in the electric cable industry

The electrical cable industry for buildings is obliged to produce cables whose sheaths are electrically insulated but also flexible enough and fire-resistant enough to maintain the electrical conductivity for as long as possible while limiting the risk of fire. Thanks to their intrinsic properties, some polymer materials meet the specifications of electrical insulation and flexibility; however, most of them are highly flammable.

For a long time the solution chosen to overcome this low fire resistance was the use of halogenated materials, capable of trapping the free radicals that are responsible for the combustion in air and the thermo-oxidation of organic materials into fuel. Unfortunately, it has been found that these additives lead to the formation of toxic products during combustion (ashes, dioxins, etc.) and alternative solutions have been proposed. Among them, the use of aluminium hydroxide ($Al(OH)_3$) appeared to be promising. Due to the heat, aluminium hydroxide produces water, which dilutes the fuel and the oxygen around the flame. On the other hand, this production of water is highly endothermic and reduces the temperature of the material during its evaporation. Unfortunately, to reach a level of efficiency in fire resistance, a composite must be more than 70 per cent aluminium hydroxide, in weight, within the polymer matrix. In these conditions, the preparation of this composite requires a lot of energy and the resulting sheath tears easily during twisting and folding.

The use of a small amount of montmorillonite (less than 3 per cent in weight) dispersed in that kind of composite allows a reduction of 55 per cent in the quantity of aluminium hydroxide required to obtain more flexible cables, having the same fire resistant qualities while consuming less energy.

8.5 Prospects

Science is at the dawn of its use of nanoparticles as dispersal agents in organic matrix-based composites. The first research work in this domain dates to the end of the 1980s and the current applications still only essentially concern nanocomposites based on smectite nanolayers. These, due to their mechanical resistance, properties as a fluid barrier and fire resistance, allow us to consider uses that are more common in industries such as aviation and aerospace (requiring light and strong materials), automotive (where the reduction in weight allows substantial reduction in fuel consumption) and packaging (where light and flexible materials are required). New revolutionary materials are about to be brought out, such as polymer-nanocomposite-based foam, which are more resistant to deformation than the common foams, are capable of competing with solid materials, and at the same time have a much lower density.

Moreover, the fast-growing production, at the industrial level, of nanoparticles and particularly of carbon nanotubes allows us to imagine the probable boom in the use of nanoparticles in organic matrices over the coming years.

Finally, the emergence of new nanoparticles such as magnetic or superconducting nanolayers, and also the huge progress that is occurring in supramolecular chemistry, open the door to the development of smart materials with outstanding properties.

Chapter 9
Nanomagnetism

Among the different fields concerned with nanotechnologies, that of magnetic particles plays an important role, particularly in biomedical applications.

9.1 Magnetism in matter

The magnetism of condensed matter is due to the magnetic moments of electrons and depends on the way that these moments combine. A material responds to the presence of an external magnetic field B_0 by a magnetization M_0, also known as magnetic moment by unit of volume ($B_0 = \mu_0 H$, where μ_0 is the magnetic permeability of the vacuum). The material magnetic properties are characterized by the magnetic susceptibility (quantity without dimension):

$$\chi = M_0/H \tag{9.1}$$

9.1.1 Diamagnetism and paramagnetism

When the magnetic moments of electrons, which bond in an orbital, compensate each other, the material is known as *diamagnetic*. For diamagnetic materials, χ is negative and very low, of the order of 10^{-5}.

When the magnetic moments of the electrons do not cancel, a group of elementary magnetic moments of module μ, independent of each other and subject to an external field B_0, tends to align according to B_0 and to lead to a macroscopic magnetization:

$$M = M_S L(x) \tag{9.2}$$

where M_S is the magnetization at the saturation, i.e. the magnetization when all the elementary moments are well aligned; $L(x) = \coth(x) - 1/x$ is the Langevin function (Figure 9.1), and $x = \mu B_0/(k_B T)$, k_B is the Boltzmann's constant and T the absolute temperature.

By developing the Langevin function at the first order at x, for a low field or a high temperature, it becomes $L(x) \sim x/3$, and consequently

$$\chi = \mu_0 n \mu^2/(3k_B T) \tag{9.3}$$

where M_S has been replaced by $n \cdot \mu$, n being the number of elementary magnetic moments by unit of volume. Equation 9.3 defines the paramagnetic Curie susceptibility.

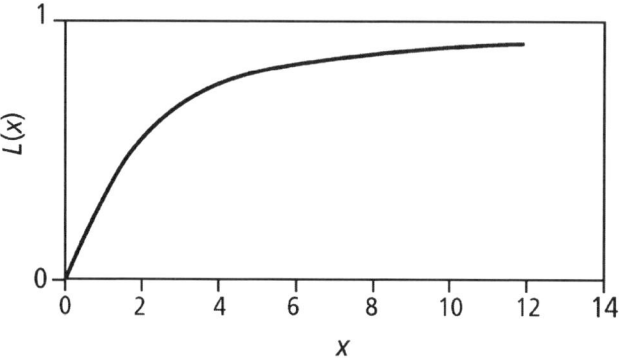

Figure 9.1   Langevin function

The assumption of independence of the elementary moments, which allows us to establish (9.3), is obviously not really realistic; on the contrary, the magnetic moments of electrons within the condensed material are in general highly coupled, via the exchange interaction, whose Hamiltonian is expressed as:

$$H = -\sum_{IJ} J_{ij} S_i S_j \tag{9.4}$$

This coupling favours the alignment of spins when the exchange integral J_{ij} is positive; while when it is negative, it favours the anti-alignment of spins, which are close to each other. The coupling described in (9.4) results from the overlap of electronic orbitals, and from the fact that wave functions associated with electrons are antisymmetric with respect to the permutation of two electrons (Pauli principle). The magnetic moments are related to the angular moments via the gyromagnetic ratio $\gamma = \mu/J$, where J is the total angular moment.

9.1.2   Ferromagnetism and Weiss domains

Pier Weiss has elaborated a ferromagnetism model based on the idea that, at the first approximation, each spin is subjected to the influence of its neighbour by being subjected to a mean field proportional to the magnetization of the material. Equation 9.3 is then modified as follows (only when $T > T_c$):

$$\chi = C/(T - T_c) \tag{9.5}$$

where T_c, known as *Curie temperature*, depends on the exchange integral and on the constant of proportionality between the mean field and the magnetization. The Curie temperature characterizes a phase transition: for $T > T_c$, the material is paramagnetic, while for $T < T_c$, the material is ferromagnetic and presents a spontaneous magnetization not equal to zero, in the absence of external field. The Curie temperature of iron is equal to 1 043 K, and for cobalt, T_c is equal to 1 394 K.

However, the Weiss theoretical model does not show an important experimental result: the existence of different domains within a macroscopic sample, each of them being characterized by a uniform magnetization, different from one domain to another. These domains (also known as *Weiss domains*) divide the volume of the sample, and the magnetic moment of this sample is nil in the absence of an external field. Consequently, the existence of these domains prevents the material from having an arbitrary magnetization direction in the absence of external anisotropy instruction, and the theoretical issue of spontaneous loss of symmetry is bypassed.

When an external field is applied, the domains whose magnetization directions are favoured regarding the required energy will have their volume increased to the detriment of those whose magnetizations point in a direction which requires more energy (see Figure 9.2).

Finally, the choice of the magnetization direction in a crystal is not free, unlike in a strictly paramagnetic model, where the elementary magnetic moments interact only with the external field. There are so-called 'easy' axes, along which the magnetization tends to align, in order to minimize the anisotropy energy of the system. The magnetization is not the result of only one parameter: it is codetermined by the magnetocristalline anisotropy energy – which itself depends on the crystal structure – and by the energy of demagnetization, which depends on the external shape of the magnetized device, and by some surface effects. The simplest model, which is particularly appropriate for cobalt and almost always valid for the contribution of the demagnetization, is the uniaxial anisotropy energy model, whose analytical form is expressed as

$$E_A = CV \sin^2 \theta \tag{9.6}$$

where θ is the angle between the magnetic moment vector and the 'easy' magnetization direction, and V is the initial volume of the crystal (C is an energy per unit volume). For crystals with more complex symmetry, the anisotropy energy has a more complex form; it is expressed as

$$E_A = CV(m_x^2 m_y^2 + m_y^2 m_z^2 + m_z^2 m_x^2) \tag{9.7}$$

where m_x, m_y, m_z are the direction cosine of the magnetization vector; x, y, z being the cubic symmetry axes of the crystal.

These domains are formed to save the magnetostatic energy (also known as dipolar energy). The magnetic field lines are always closed loops – as described by Maxwell laws ($\nabla B = 0$) – and a uniform magnetization medium can close its self-generated field lines only outside the medium, which requires a quantity of energy $B^2/(2\mu_0)$ by unit of volume. However, two adjacent domains whose magnetizations are opposite require magnetic field lines shorter than the body itself (see Figure 9.3).

However, the formation of the wall that separates two domains and within which the elementary magnetic moment vectors turn 90° (or 180° depending on the case) also requires energy, which can be quantized with the help of a parameter σ representing the formation energy of the wall by unit of surface. If this energy, for a simple wall that partitions a sphere of radius R in 2, i.e. $\pi R^2 \sigma$, is compared to the dipolar energy of a sphere of uniform magnetization M, i.e. $\mu_0 \pi M^2 R^2/9$, we can see that the formation energy of the wall is smaller than the dipolar energy if $R > 9\sigma/(\mu_0 M^2)$,

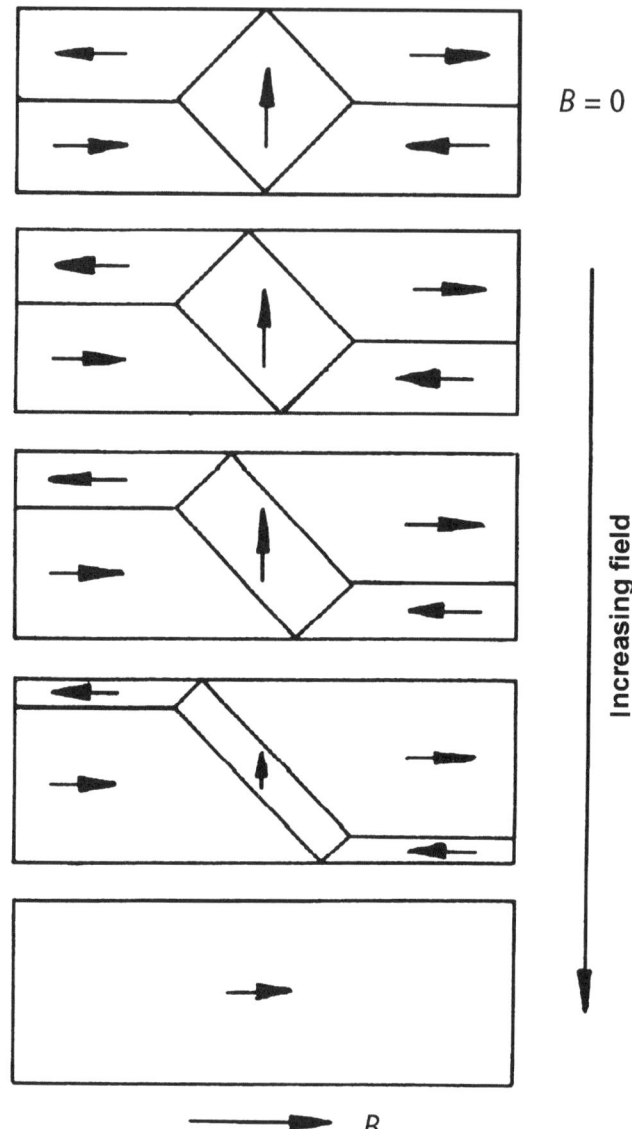

*Figure 9.2   Adaptation of Weiss domains according to the increase in the external
    magnetic field*

i.e. $R > 10^{-7}$ when $\sigma \approx 10^{-2}$ J/m^2 and $\mu_0 M \approx 1$ T (typical value of magnetite); here it is the surface/volume ratio that is at stake. For a radius smaller than this limit, there will not be any formation of domains, and the crystals will be uniform (single-domain).

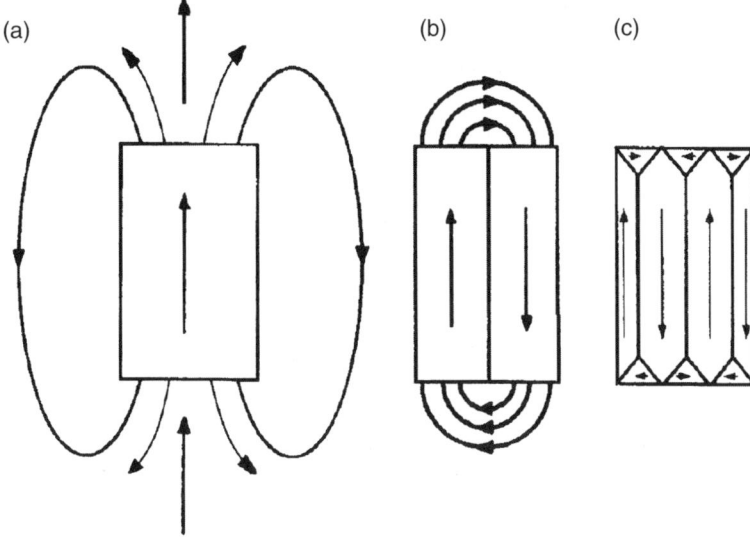

(a) (b) (c)

*Figure 9.3   Magnetic field lines of a bar magnet: the figure illustrates the same
bar magnet, containing 1, 2 and 10 domains. It shows the field lines
disappearing outside the bar*

9.1.3   Superparamagnetism

For nanoparticles in the absence of an external field, an abrupt change of symmetry is expected. The magnetic-moment vector 'chooses' between the different 'easy' directions, which are all energetically equivalent without any external indications. However, the anisotropy energy (Equations 9.6 and 9.7) is proportional to the particle's volume. The energy barrier to go from an easy magnetization direction to another – i.e. in the case of a single-axis anisotropy, from one side to the other on the same axis – also decreases with the size of the crystal, and temporal fluctuations can then appear during which the magnetic-moment vector jumps from an easy direction to another. The correlation time that characterizes these random fluctuations is known as the Néel time (τ_N), and is an exponential function of the anisotropy energy:

$$\tau_N \alpha \sqrt{\frac{k_B T}{KV}} \cdot (e^{\frac{KV}{k_B T}}) \tag{9.8}$$

where we assume that $E_A = KV$. This time is roughly of the order of 10 ns for a magnetite crystal of 10 nm diameter at ambient temperature. It is short enough for an effective measure of magnetization, which gives zero in the absence of an external field. This solves the issue of a sudden break in the symmetry, by a temporal average rather than a spatial average, as in the case of Weiss domains. When the ferromagnetic nanoparticles are subjected to an external field, the fluctuations do not disappear, but the magnetic-moment vector spends more time in the easy directions close to the direction of the external field, leading to a magnetization proportional to this external field. Then, the behaviour of these nanoparticles is comparable to the behaviour of paramagnetic bodies, apart from the fact that their susceptibility is higher because of the collective behaviour of the electrons imposed by their exchange interactions. This justifies the designation of superparamagnetic given to them: paramagnetic susceptibilities are of the order of 10^{-4}, while superparamagnetic susceptibilities are often 100 times greater.

9.1.4   Antiferromagnetism

As mentioned above, the exchange integral that plays a part in (9.4) can be negative; the neighbouring spins then minimize their interaction energy by opposing each other. The lattice can then be considered as formed of two sub-lattices which interpenetrate, and whose magnetizations (which have the same modulus) are situated in opposite directions. Such materials are known as *antiferromagnetic*. They are also characterized by a phase transition. When the temperature is higher than the Néel temperature (T_N), i.e. the critical temperature, then the material is known as *paramagnetic*, while for $T < T_N$, the material is antiferromagnetic.

When an antiferromagnetic material is subjected to an external magnetic field, it becomes magnetized due to two effects: on the one hand, the two sub-lattices twist and their respective magnetizations are not strictly parallel any more, and a resulting magnetization appears in a direction perpendicular to the easy magnetization directions of the sub-lattices; on the other hand, and particularly for small crystals, a certain number of knots of the lattice are without any moments under the effect of an imperfect crystallization. These defects show uncompensated moments, which lead to a magnetization component parallel to the easy magnetization direction. Due to the superparamagnetic process described by Louis Néel (Figure 9.4), this magnetization component jumps from one side to the other of the easy magnetization axis. The susceptibility of antiferromagnetic material is around 10^{-3}, which is much lower than the susceptibility of superparamagnetic materials (characterized by susceptibility of the order of a ferromagnetic material).

9.2   Superparamagnetic colloids

9.2.1   Properties

Superparamagnetic colloids, such as ferrites, are suspensions of nanometric single crystals each composed of a single ferrimagnetic or ferromagnetic domain. Their potential application fields are wide and extend from mechanics to medicine. In the biomedical applications, these systems are used as diagnosis and therapy agents: in magnetic resonance imaging (MRI), they act as contrast agents, and in cancer treatment they absorb the high-frequency electromagnetic waves that can induce a localized hyperthermia leading to the selective destruction of tumour cells.

Ferrites, which are metal oxide particles, are compounds with interesting magnetic properties. They have a general formula (Fe_2O_3, MO), where M symbolizes a bivalent metal ion. Magnetite, which is a ferrite where the bivalent ion is an iron, is the most

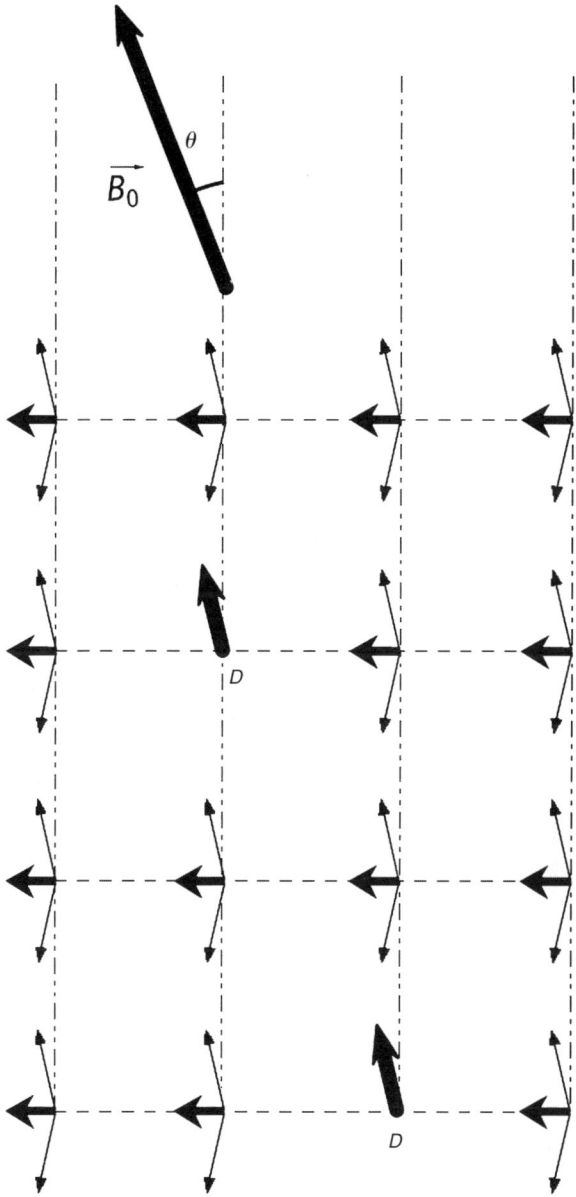

Figure 9.4   *Antiferromagnetic crystal model. The dashed lines represent the one direction of easy magnetization; each knot has elementary opposite moments. Under the effect of the field B_0, the two sub-lattices deform, leading to elementary moments represented by the horizontal arrows. The defects (pointed out by the D characters) produce a contribution to the magnetization parallel to the easy magnetization direction*

used of this group of compounds. In medicine, metal particles of iron or cobalt are not used because of the oxidation phenomenon that occurs in aqueous medium and leads to the formation of nonmagnetic oxide.

A colloid solution is a stable dispersion of particles in a continuous phase (liquid, solid, vapour). The size of colloidal particles can vary from a few nanometres to a few dozen micrometres or more.

These superparamagnetic colloids, also known as *ferrofluids*, are composed of magnetic grains stabilized in stable suspension form because of repulsive forces that prevent the grain from agglomerating. In fact, magnetic attractions between grains lead to their agglomeration, and consequently the destabilization of the system. This phenomenon can be cancelled by the surface adsorption of stabilizer grains that induce repulsions, preventing the grains from agglomerating.

These colloids are classified in two different groups according to the type of the stabilizer:

- UMF (uncoated magnetic ferrofluids): the crystals are stabilized because of a surface charge (see Figure 9.5);
- SMF (surface magnetic ferrofluids): the magnetic grains are stabilized by steric repulsion due to the presence of long chains adsorbed on the surface of the grains (see Figure 9.6).

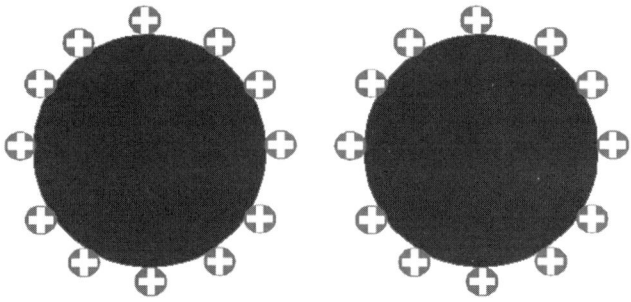

Figure 9.5   Particles stabilized by surface charge

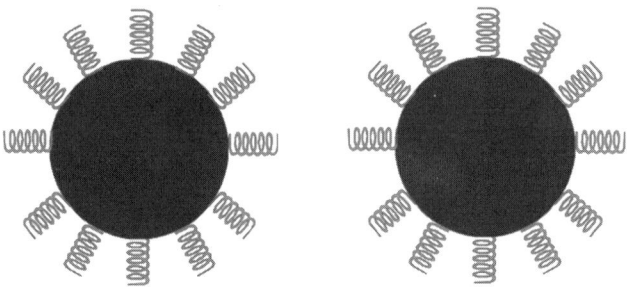

Figure 9.6   Particles stabilized by a steric coating

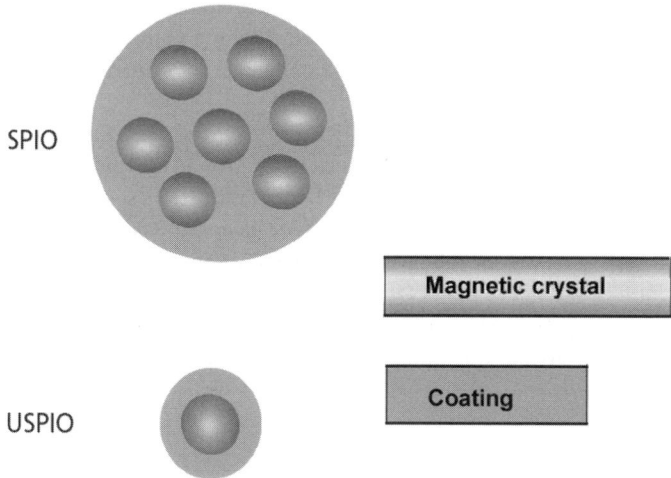

Figure 9.7   Illustrations of SPIO and USPIO

In order to prevent any terminological ambiguities, the term *grain* is reserved for the magnetic heart of the compound, while the term *particle* refers to the magnetic grain and its coating (ionic or steric coating).

The aggregation state of the particles depends on the type of synthesis of these ferrofluids. Two groups of nanometric particles can be distinguished here:

- the SPIO (small particles of iron oxide), where each particle contains many magnetic grains;
- the USPIO (ultra-small particles of iron oxide), characterized by a single crystal per particle.

9.2.2   Synthesis

Molday's synthesis is the starting point of most syntheses of ferrofluids based on the coprecipitation of ferrous and ferric ions in an alkaline medium. An alternative synthesis is based on a ferrous salt solution with a basic pH in the presence of an oxidizing agent. This protocol is modified, depending on the way the grains are stabilized (ionic or steric).

Depending on the application of these ferrofluids, many types of steric protection are used. In medical applications, biocompatibility and biodegradability will impose the choice of the stabilizer polymer; synthetic polymers such as polyethylene glycol (PEG) have been used to produce contrast agents for imaging. Magnetic balls coated by synthetic polymers have also been used for cell separation, for the study of inter-actions between proteins and for the transport of active ingredients. Magnetic grains coated with proteins, such as human serum albumin (HAS), have been developed for the targeting of medicines.

Some magnetic particles have also been given an antibodies graft to their surface in order to target a specific receptor expressed during a pathology, and to allow the internalization of magnetic particles by the cells and consequently a contrast agent specific to this pathology.

In the group of contrast agents for MRI of the SPIO-type, Endorem®(Guerbet, Aulnay-sous-Bois, France) – magnetite particles of around 200 nm in diameter coated with dextran – is used for the diagnosis of liver tumours. Resovist® (Shering, AG, Berlin, Germany) – also known as SH U 555A – is a suspension of magnetic grains coated with carboxy-dextran. The particles have a diameter of around 60 nm with a magnetic grain of 4.2 nm in diameter.

In the USPIO group, Sinerem® (Guerbet, Aulnay-sous-Bois, France) is used, obtained from size division of Endorem®. It is a suspension of particles of around 20 nm in diameter.

9.2.3   Magnetoliposomes

A liposome is a microscopic blister the size of which can vary from a few dozen nanometres to a few micrometres. The membrane of these blisters is composed of a phospholipidic double layer with a structure similar to living cells. These systems have the advantage of being able to merge with cell membranes, which makes them capable of transporting active ingredients. In the cosmetics field, the use of liposomes improves the effectiveness of creams by slowing down the diffusion of the active ingredients that they encapsulate.

Magnetoliposomes are magnetite grains of a few nanometres' diameter coated with a phospholipidic double layer. Their properties differ according to the types of phospholipid contained in the membrane. The surfaces of these systems are appropriate to favour the targeting of a pathology and for magnetic detection, thanks to the properties of magnetite grains.

9.2.4   Characterization of superparamagnetic colloids

The effectiveness of superparamagnetic colloids in MRI is quantized by the measure of the acceleration of protons' relaxation in the solvent (often water). The relaxivity is the increase in the relaxation speed due to the presence of a millimole of iron per litre.

In fact, ferrite particles in suspension speed up the longitudinal and transversal nuclear magnetic relaxations of protons of water. The increase in relaxation speeds depends on the intrinsic properties of particles, i.e. the size of the magnetic grain, its composition, its magnetic properties and the nature of its coating as well.

It is consequently important to characterize the particles well in order to anticipate their effectiveness as contrast agents in medical imaging.

The composition of the crystal will be worked out by X-ray diffraction, and its magnetic properties by Mössbauer magnetometry measures and nuclear magnetic relaxation dispersion profiles (NMRD).

Magnetometry graphs represent the magnetization measure of magnetic grains as a function of the applied magnetic field. For superparamagnetic fluids, these curves do not show any remanence and the magnetization saturation is very high. The smoothing of these curves by a Langevin function allows the deduction of the specific magnetization as well as the mean size of the magnetic grain.

Mössbauer spectroscopy allows the oxidation state of iron ions within the crystal to be shown with certainty and allows the order of magnitude of the Néel relaxation time to be characterized.

NMRD profiles give the longitudinal relaxation speed of protons in water as a function of the static magnetic field in which these protons relax. The smoothing of these profiles by the model of superparamagnetic relaxation shows the size of the particles and their specific magnetization as well as the order of magnitude of the Néel relaxation time.

Transmission electronic microscopy (TEM) is used to measure the size of the magnetic grain. This technique also allows an assessment of the size dispersion of the magnetic core to be made, provided that the measure is done on a great number of particles. Difficulties can arise using TEM due to the possibility of agglomeration of the grains during the preparation of the sample. In fact, the measure is done on a dried sample and it is the change from the colloidal to the solid state that can lead to the formation of aggregates.

Photonic correlation spectroscopy (PCS) measures the hydrodynamic diameter of the entire particle (grain and coating). The measure of the light intensity, diffused by the colloid irradiated by a laser beam, allows us to measure the particle's coefficient of diffusion, and, thanks to the Stokes–Einstein equation, to deduce the size of the particles in suspension.

The sizes obtained by the different methods mentioned above are in perfect accordance when the analysed sample contains particles of a single size (the sample is then known as *monodispersed*), but the measures diverge when this condition is not fulfilled.

9.3   Nanomagnets in thermotherapy

We have already discussed the use of nanomagnets, built for specific receptors, to allow a very selective detection of pathologies thanks to magnetic resonance imaging. Magnetic nanoparticles help, therefore, in determining the diagnosis. They can also be used as therapeutic agents. In fact, a new way of treatment of tumours can rapidly be developed, in addition to the usual methods of surgery, radiotherapy and chemotherapy: thermotherapy.

9.3.1   Heating tumours to destroy them

Thermotherapy consists in irradiating the patient with non-ionizing magnetic waves, in order to increase the temperature within and around the tumour. Tumorous cells are more sensitive to an increase in energy than healthy cells. A selective destruction of a pathological area can then be expected. The therapeutic effect is achieved when the temperature exceeds 43 °C at the tumour. However, the use of this technique is limited because of the bad delimitation of the volume subjected to the hyperthermia. In fact, the thermotherapy destroys a tumour but it can also seriously damage healthy cells.

Currently, the delimitation of the zone to heat is obtained using a group of dipolar antennae placed around the patient. The targeting of the zone to be treated has first been improved using a small antenna implanted in the pathological tissue – implantation that is not without risk for the patient. Another technique consists in using ferromagnetic implants. They pick up the electromagnetic wave and convert it into heat via a principle similar to the one responsible for the increase in temperature in electrical supply transformers. However, these macroscopic sensors are not the most effective heaters to convert the electromagnetic energy into thermal energy. According to the first clinical tests, the combined use of thermotherapy and radiotherapy can, in some cases, double the survival rate after two years compared with radiotherapy treatment alone.

Actually, the best converter of electromagnetic energy into heat is very likely a material composed of superparamagnetic nanomagnets. These nanomagnets, mixed into the paints of fighter planes, make them invisible to radar systems, whose emissions get absorbed by the surface of the planes.

By grafting to nanoparticles a molecular entity capable of targeting a specific given pathology, it is possible to concentrate them in a noninvasive way in the tissue to be treated, and consequently to selectively heat the pathogenic cells.

Why thermotherapy destroys tumorous cells rather than healthy cells

Cancerous cells develop very quickly and consequently require a huge amount of oxygen and nutriments. To fulfil this requirement, blood vessels will multiply. However this anarchical proliferation leads to the appearance of loops and venous obstructions. The final result will be a very bad irrigation of the tumour despite extensive vascularization. If an electromagnetic wave brings energy to tumorous cells, its dissipation will be very difficult because of the bad irrigation. On the contrary, in healthy cells, the blood flow allows an easy evacuation of the calories brought. The destruction of pathologic cells comes from a fluxing and an alteration of the cells' membrane, as well as the damage caused on the cytoplasmic backbones and the cores. Cancerous cells are also more sensitive to hyperthermia because of their high acidity. This relatively low pH is another consequence of the bad irrigation; in fact it is the result of the difficulty of eliminating the waste inherent in anaerobic metabolisms. The combined effects of the increase in temperature and the acidity irreversibly alter the cellular proteins.

9.3.2 Absorption of radiofrequency waves by nanomagnets

How hyperthermia improves the effectiveness of the treatment by ionizing radiation

Ionizing radiations have the effect of breaking the chemical bonds to produce unstable entities known as *free radicals*. These free radicals cause serious damage to DNA molecules, which leads to the destruction of cells. The effectiveness of this process decreases by a factor of 3 when there is a lack of oxygen in the cell. The hyperthermia increases the blood circulation and allows a better oxygen supply in the tumorous cells, which induces a higher sensibility to ionizing radiations.

The conversion of electromagnetic energy into heat requires a dissipative mechanism, i.e. a process comparable to friction forces. In the case of magnetic crystals, the physical parameter that allows the characterization of the effectiveness of this dissipative phenomenon is the relaxation time of the magnetization of the material.

To understand better the phenomenon of magnetic relaxation in the process of conversion of the magnetic energy into calories, let us consider a sample composed of a group of nanomagnets, in a context where no magnetic field is applied. The magnetization of this sample is then nil. What would happen if, at time t_0, all the nanomagnets were subjected to the effect of a magnetic field B_0? Following the example of the needle of a compass pointing to the north pole, the magnetic moments of crystals tend to point to the direction of the magnetic field B_0. This preferred direction leads to the appearance of a global magnetization M_0, laid down by the laws of thermodynamics. This magnetization – as long as the magnetic field remains low – is given by (9.1), which defines the magnetic susceptibility of the material.

The conditions that govern the irradiation of nanomagnets by oscillating magnetic fields are generally such that (9.1) is applicable.

The magnetization process of the sample is however not instantaneous, but is characterized by a relaxation time τ (see Figure 9.8). The temporal evolution of the magnetization $M(t)$ is ruled by the equation

$$\frac{d(M(t))}{dt} = -\frac{1}{\tau}(M_0 - M(t)) \tag{9.9}$$

The solution to this differential equation is given by

$$M(t) = M_0 \cdot \left(1 - e^{\frac{-(t-t_0)}{\tau}}\right) = \chi_0 \cdot H \cdot \left(1 - e^{\frac{-(t-t_0)}{\tau}}\right) \tag{9.10}$$

The process described above is ruled by two different phenomena. The first is a purely mechanical process: the rotating movement of the entire crystal subjected to thermal agitation. It is described by τ_B, the Brown relaxation time (see Figure 9.8).

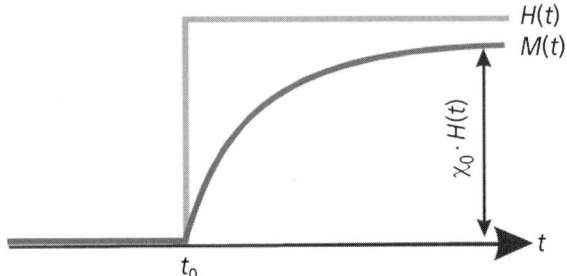

Figure 9.8   Magnetization of nanomagnets: evolution of the magnetization after applying a magnetic field to a group of nanomagnets. The evolution of M(t) is given by (9.10)

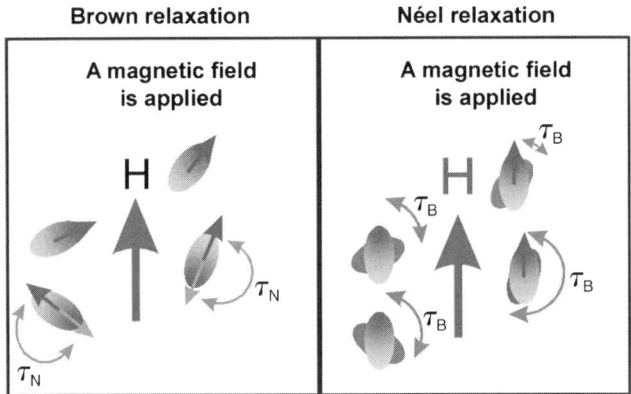

Figure 9.9   Brown and Néel relaxation process

The second one is the Néel relaxation, which characterizes the evolution of the orientation of the magnetic moment in relation to the crystal. Generally, this orientation is marked out with respect to one or many specific crystalline axes known as *easy axes*. These axes correspond to directions for which the interaction energy between the magnetic moment and the electronic orbitals of the crystal is minimal (see Figure 9.9).

The relaxation speeds, which are proportional to the inverse of the relaxation times, are additive, and the return to the magnetization equilibrium is characterized by a time constant τ given by

$$1/\tau = 1/\tau_B + 1/\tau_N \tag{9.11}$$

It is therefore the fastest mechanism that dominates the relaxation. If the nanocrystal is in powder form or is attached to a rigid support, only the Néel relaxation time plays a part in the magnetic relaxation process.

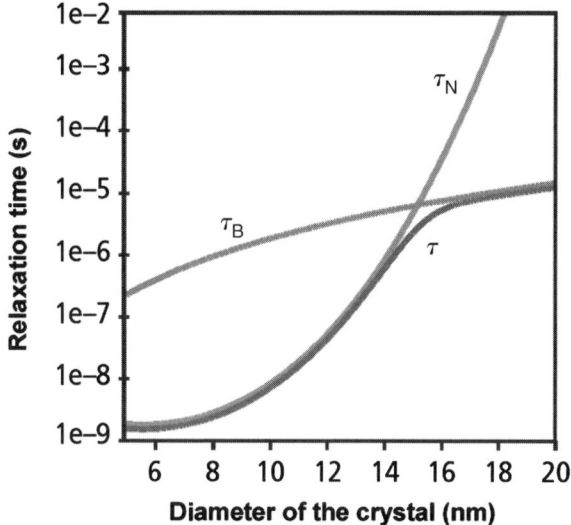

Figure 9.10   Magnetic relaxation time: illustration of the magnetic relaxation time τ and its two components, Néel (τ_N) and Brown (τ_B) relaxation times, as a function of the size of the crystal, for a suspension of magnetite nanomagnets in water

In colloidal solution, the two relaxation mechanisms coexist. The Brown relaxation time is given by the Stokes–Einstein relation, which predicts the linearity of this relaxation time with the volume of the magnetized particle. The Néel relaxation time presents a dependency regarding a much bigger volume, as shown in (9.8): due to the presence of the volume of the particle in the exponent of the exponential, the Néel time increases very rapidly with the diameter (see Figure 9.10).

Regarding the best-known magnetic iron oxide, magnetite (Fe_3O_4), the Néel relaxation dominates for a diameter of less than 14 nm. On the contrary, for magnetite particles of a diameter greater than 14 nm, the magnetic relaxation is ruled almost exclusively by the Brown relaxation.

However, the behaviour of nanomagnets is different depending on whether they are in solution or rigidly attached to the wall of a cell.

The magnetic relaxation plays an essential role in the conversion of electromagnetic energy into heat. To obtain hyperthermia, the sample containing the nanomagnets is subjected to an oscillating magnetic field. At low frequency, i.e. when the oscillation period is longer than the relaxation time, the magnetization vector of the sample follows accurately the oscillation of the magnetic field. The energy dissipation being proportional to the speed of magnetization variation, it consequently increases with the oscillation frequency of the magnetic field.

At high frequency, i.e. for an oscillation period much smaller than the magnetic relaxation time, the magnetization vector of the sample does not have time to 'respond' to the variation of the magnetic field. The magnetic vector then remains static and

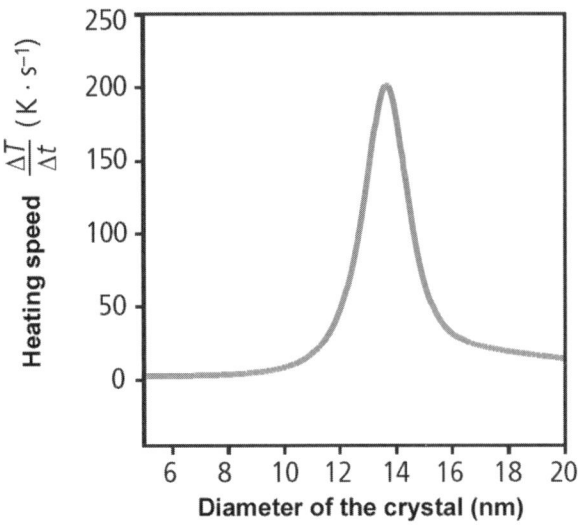

Figure 9.11   *Heating speed vs. size of crystal: graph showing the dependency of the heating speed regarding the crystal's size in a sample containing nanomagnets of magnetite in a volumic proportion of 7.1 per cent, subjected to a field of 2.4×10^5 A/m*

very little energy is dissipated. The energy dissipation will then be maximum for an intermediary frequency for which the magnetization relaxation time will be at the same order of magnitude as the oscillation period of the field ($\omega\tau = 1$).

For a frequency of 100 kHz, the magnetic relaxation time allowing the maximum caloric dissipation is of 1.6×10^{-6} s, which corresponds to a magnetite nanocrystal of around 14 nm diameter.

The dependence on crystal size of the heating speed of a suspension of magnetite nanomagnets shows a narrow range (see Figure 9.11): the most efficient particles that induce hyperthermia at a frequency of 100 kHz will then have a diameter of 14 nm. Furthermore, note the swift decrease in the efficiency of the thermal effect when the width of the distribution of the particle size increases. The decrease in performance of the nanomagnets with the increase in the polydispersiveness in size has besides been confirmed by numerical simulations.

Two conditions are essential in order to use magnetic fluids in thermotherapy: first, particles with a specific diameter and a narrow size distribution must be produced; then we must be able to graft to these nanomagnets some molecular or biological entities that target the specific receptors to the pathology to be treated.

9.3.3   Results

Classical hyperthermia, without the intervention of magnetic particles, uses high-frequency waves (between 100 and 400 MHz). For these frequencies, the thermal effect is induced by the electrical component of the electromagnetic wave, which generates ionic currents. The difference in the electrical conductibility of tissues then leads inevitably to a bad definition of the zone subjected to the hyperthermia, even if the irradiation is well delimited by an appropriate system of transmitting antennae.

The use of nanomagnets can resolve the issue of the selection of the zone to heat. In fact, in this case the dissipated heat is directly proportional to the concentration of magnetic nanocrystals. The heated zone will then be the reflection of the bioselectivity of the nanomagnets for the pathology to be treated. The frequency of the variation of the magnetic field used (between 100 and 300 kHz) will be a thousand times lower than the one used in a classical thermotherapy irradiation. This allows a heat dissipation caused exclusively by the interaction between magnetic field and magnetic nanocrystals.

In fact, the idea of using ferrimagnetic or ferromagnetic particles to catch an electromagnetic wave and to transform it into heat for hyperthermia dates from 1960. However, it was only in 1963 that Jordan *et al.* showed experimentally that the use of superparamagnetic nanocrystals allows us to obtain a speed of conversion of magnetic energy into heat much higher than when ferromagnetic crystals are used. Six years later, this technique was used with success to stop the growth of mammary-carcinoma-type tumours implanted in mice. The cells were charged with magnetite particles of a mean diameter equal to 13.1 nm, and coated with aminosilane. Some preliminary *in vitro* tests showed that cancerous cells catch 10 times more of this kind of nanomagnet than the healthy cells.

Since 2000, a prototype appliance has been in development at the Charity Clinic of Berlin and some clinical tests on cerebral tumours have been conducted. With this appliance, tumorous tissues in the patient can be subjected to a magnetic field oscillating at 100 kHz. As in the case of the classical treatment by high-frequency waves, the treatment is combined with radiotherapy. The magnetic fluid is brought to the tumour level by the use of a stereotactic injection. The injected amount is of 15 mg/g of tumour. The use of smart nanomagnets, capable of localizing pathology zones after a bolus injection will depend on future progress on the molecular targeting.

To conclude, thermotherapy induced by magnetic fluids is certainly a promising technique. However, some effort is still needed in order to improve the efficiency and the bioselectivity of the nanomagnets used.

9.4   Biomagnetism

Industrial applications of magnetic nanoparticles are various: from the storing of data to the increase in contrast in nuclear magnetic resonance imaging to the technology of radar stealth. However, the synthesis of these particles is not always easy and repeatable. Therefore, it is surprising to note that some living organisms have made them for millions of years, taking advantage of their extraordinary properties. Similarly, the presence of endogenous superparamagnetic particles in certain organs of the human body have unexpected consequences for the contrast obtained in medical imaging. The magnetism of these particles synthesized *in vivo* could even be used for diagnosis purposes.

9.4.1   Iron in biology

Iron is the second most abundant metal on the Earth's crust, after aluminium. Many living beings, from bacteria to the elephant, use it for its different chemical properties. Iron allows the transport, via haemoglobin, of the oxygen we breathe as well as the fixation of nitrogen, and iron is also involved in most photosynthesis reactions. All magnetic nanoparticles fabricated *in vivo* are iron oxides, except for greigite (Fe_3S_4) produced by some bacteria. Until now, the oxides found are ferrihydrite ($5Fe_2O_3$–$9H_2O$), goethite (α-FeOOH), lepidocrocite (γ-FeOOH) and magnetite (Fe_3O_4). Table 9.1 summarizes the crystalline and magnetic properties of these different compounds.

9.4.2   Molluscs

Magnetite can be found on the small teeth that line the surface of chiton radulae (see Figure 9.12). The radula is like the 'tongue' of this mollusc, which it uses to pull up seaweed from rocks on the sea bottom. The magnetite coating of the teeth, of a maximum thickness of 10 μm, is to reinforce and slow down their wear. In fact, the formation of magnetite crystals on the chiton radulae is quite complex: first, ferrihydrite

Table 9.1   Crystalline and magnetic properties

Name	Crystalline system	Type of magnetism
Ferrihydrite	hexagonal	antiferromagnetic
Lepidocrocite	orthorhombic-disphenoidal	antiferromagnetic
Goethite	orthorhombic-disphenoidal	antiferromagnetic
Magnetite	cubic	ferrimagnetic

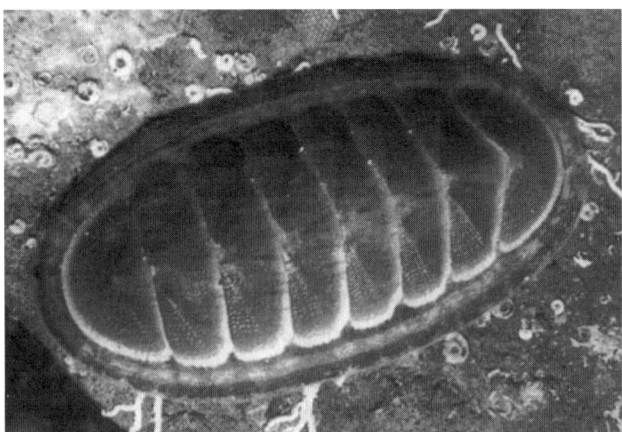

Figure 9.12   Chiton

crystals grow in an organic matrix at the teeth's surface, to be then transformed into magnetite. Therefore, immature teeth have a red colour (due to the ferrihydrite) while the mature teeth are dark (because of the magnetite). Hydrated iron oxides can also be found in the coating of the teeth of the chiton radula, as lepidocrotite and goethite. The latter is therefore the principal component of the coating of the teeth of the radula of another mollusc: the patella. In these molluscs, iron oxides are not used for their magnetic properties, but for their hardness and their wear resistance.

9.4.3   Magnetotactic bacteria

Magnetotaxis is the ability to find one's bearings and to move according to the direction of a magnetic field. In this way, magnetotactic bacteria move according to Earth's magnetic field lines in order to find the sediment essential to their survival. Discovered in 1975 by Blakemore, they use magnetite crystals (or sometimes greigite) to line up according Earth's magnetic field. These crystals range in size from 30 to 100 nm, are coated with a phospholipidic membrane to form what is known as a magnetosome. The different magnetosomes of similar bacteria are often arranged in line to form a real compass needle on the length of the bacterium, as illustrated in Figure 9.13.

Magnetite crystals of magnetosomes are of a slightly smaller size than that of a magnetic monodomain of magnetite. Consequently, their magnetic moments are very high. However, their size is too big for some superparamagnetism to be noticed: the Néel relaxation time of such crystals is of the order of 10^{131} years. These crystals then have an optimal size: their magnetic moments are maximal, but do not change according to the different anisotropy directions, since the Néel relaxation time is very high. They are then a real 'nanocompass'. The process of orientation and movement along the magnetic-field lines is very easy: the magnetosomes chain of the bacterium can be represented by a permanent magnetic dipole. This dipole is directed towards either the front of the bacterium or the back of the bacterium, where its locomotion organ, the flagellum, is located. In the northern hemisphere, Earth's magnetic field has a vertical component directed towards the bottom. To survive and thrive, bacteria must be close to the sea bottom, at the limit between the water and sediments. The

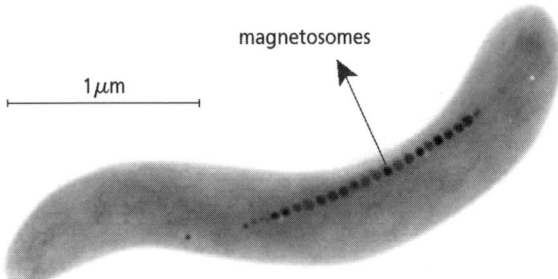

Figure 9.13   Transmission electronic microscopy picture of a magnetotactic bacterium [Source: R. Frankel]

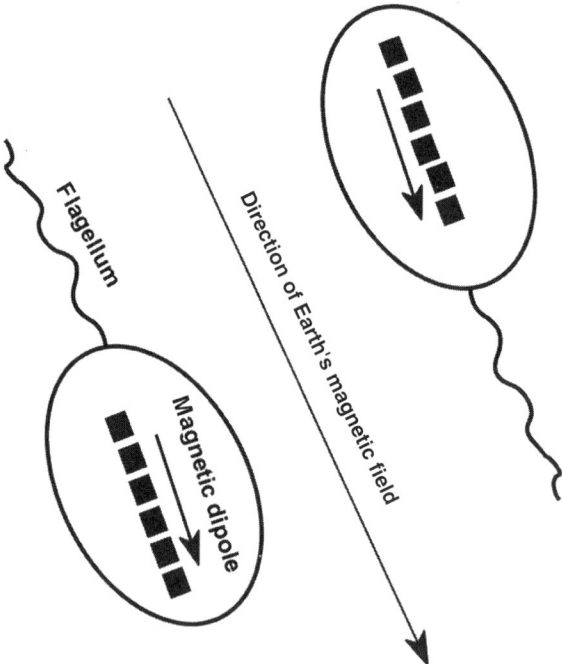

Figure 9.14   Displacement of magnetotactic bacteria in the northern hemisphere

dipole of the bacteria lines up with the magnetic field to minimize the interaction energy with the magnetic field ($E_m = -\mu \cdot B$) (see Figure 9.14).

When the bacteria activate their flagellum to move, those whose dipoles point towards the front naturally head for the bottom, while those whose dipoles point towards the back head to the surface and die. Consequently, in the northern hemisphere, most of the magnetotactic bacteria have a dipole moment pointing towards the front; they are known as *north-seeking bacteria* (NS), while the southern hemisphere is stocked with *south-seeking bacteria* (SS), whose dipole points in the opposite direction, as illustrated in Figure 9.14. At the equator, where the vertical component of the magnetic field is nil, bacteria NS and SS are in equal proportion. When the direction of the magnetic field is experimentally inverted, the bacterium makes a U-turn to align its dipole moment with the new field orientation.

A simple calculus shows that the magnetic moment of the particles (6×10^{-17} J/T) is enough to produce the alignment of the bacterium with the direction of a magnetic field of 50 μT (of the order of Earth's magnetic field). In fact, at ambient temperature, the coupling energy (E_m) is of the same order as the thermal agitation ($k_B T$). The latter is therefore not enough to 'erase' the orientation of the bacteria: $E_m = 3 \times 10^{-21}$ J, and $k_B T = 4 \times 10^{-21}$ J.

The use of magnetosomes by magnetotactic bacteria can be described as 'passive' navigation. In fact, the alignment process is purely physical – it is similar to the orientation of iron fillings according to the magnetic field lines of a magnet – and in the bacterium there is no magnetic receiver capable of interpreting the information concerning this field.

9.4.4 *The magnetic navigation of animals*

Small amount of magnetites have been found in a wide variety of animals: bees, ants, termites, rainbow trout, salmon, tritons, homing pigeons – and humans. The particles of magnetite, which are sometimes agglomerated, have variable sizes: from 1 to 200 nm depending on the case. Consequently some are superparamagnetic. The passive orientation of an animal in Earth's magnetic field – similar to the one observed for magnetotactic bacteria – is impossible: it would require a huge amount of magnetite. In order to use the information concerning Earth's magnetic field (its intensity and/or its direction), the animal must have a magnetoreceiver. This would transform the magnetic information into a nervous stimulus, interpretable by the brain. It would therefore allow migratory animals to find their bearings in the dark or with a cloudy sky, when no information on the position of the sun is available. A magnetoreceiver candidate has been identified for the rainbow trout at the olfactory epithelium level. However, for the other animals and for man, the magnetoreceivers have not yet been localized. Until this accurate mechanism of magnetoreception has been discovered, it will be difficult to prove that the magnetite present in animals is actually used for their orientation.

9.4.5 *Ferritin*

In the human body, 70 per cent of the total iron content is present in haemoglobin form, while 20 per cent is stored to be used, for instance, for the fabrication of future haemoglobin. The transport and the storage of iron in nontoxic crystalline form is carried out by transferrin (for transport) and by ferritin and haemosiderin (for storage). The real warehouse of iron, ferritin is present in many animals and vegetable species in many structural variants, and among humans it is principally found in the liver, the spleen, the pancreas, and in the extrapyramidal nucleus of the brain. It has the shape of a hollow shell of 8 nm of inner diameter and 13 nm of outer diameter. In the inner cavity there is a ferrihydrite crystal, which is a hydrated iron oxide. Up to 4 500 atoms of iron can thereby be stored by the protein. The protein when empty of iron, also known as apoferritin, is composed of 24 proteinic subunits of H-type (heavy) or L-type (light). The H-units differ from the L-units by their size and their composition.

If the apoferritin molecule comes from the liver, it will be composed of 3 H-units and 21 L-units; if it is located in a red corpuscle, it will be composed of 20 H-units and 4 L-units, etc. At the intersections of the units noticeable ducts leading to the inside of the protein are visible: 8 hydrophilic ducts located at the intersection of 3 units, and 6 hydrophobic ducts localized at the intersection of 4 units.

The ferritin is associated with different pathologies. Patients affected by haemochromatosis and thalassaemia present an excess of ferritin in the liver and

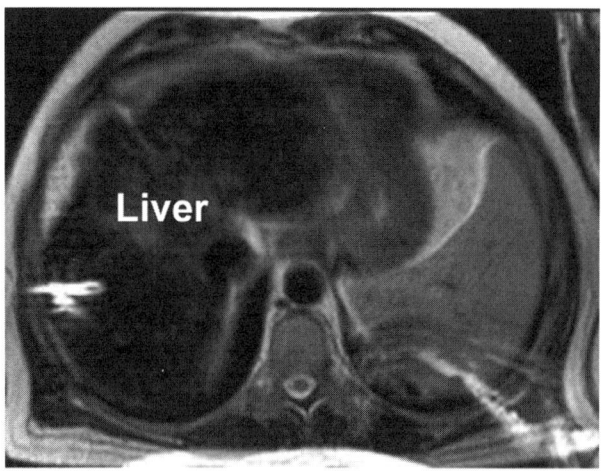

*Figure 9.15   MRI T₂-weighted picture of a patient presenting an excess of hepatic
iron: the liver is noticeably darkened*

the spleen, due to excessive absorption of dietary iron or due to iron overload as
a consequence of multiple transfusions. In the same way, too low a concentration
of ferritin in brain cells of Parkinson patients could lead to the appearance of free
radicals, which are likely to cause a deterioration of the neighbouring tissues.

The core of the antiferromagnetic ferrihydrite of ferritin is superparamagnetic,
due to its small size. In fact, the Néel relaxation time of a ferrihydrite crystal of 7 nm
in diameter is approximately 0.1 ns. The magnetic properties of the ferritin have an
unexpected effect on the contrast obtained by magnetic resonance imaging (MRI):
organs containing a great quantity of ferritin, such as the liver and some of the nuclei
of the brain, appear darker on T_2-weighted images, as illustrated in Figure 9.15. The
protein could then be described as an endogenous contrast agent. A contrast agent
was developed in 1994 on the basis of an apoferritin shell in which a magnetite crystal
had been grown. Different groups of researchers have tried to use the contrast induced
by the ferritin in MRI in order to dose, in a noninvasive way, the iron contained in
organs. However, none of the developed techniques measured the concentration of
iron as accurately as using a tissue puncture of the organ concerned.

Note that the magnetic properties of ferritin are used to measure the concentration
of iron by techniques that are purely magnetometric. In this case, a superconducting
quantum interference device (SQUID) suceptometre is used.

Chapter 10
Nanotechnologies in perspective

Many fundamental aspects of nanotechnologies remain to be studied. However, as many of the chapters in this book illustrate, applications do not wait. In 2004, around 1 500 companies worldwide (most of them startup) actively worked in research and development on nanotechnologies. At the same time, more than 300 companies (often small-scale enterprises) were producing nanomaterials. Some of them used their own product without selling it outside. At the beginning of 2006, the Woodrow Wilson International Center for Scholars, based in Washington, DC, identified a list of 190 products whose development required nanotechnology or are built with nanomaterials. Some products represent a real technology revolution (material reinforced by carbon nanotubes) while others raise more problems and issues than solutions (food, cosmetics and beauty care). Products labelled 'nano' are separated into five main groups: (1) cosmetics/healthcare/food, (2) electronics, (3) air purifiers/deodorants/disinfectant, (4) textiles, (5) reinforcement of materials/surface treatment. These are summarized along with country of origin in Table 10.1.

As explained in the first chapter, possible applications of nanotechnologies are not restricted to the fields mentioned above. The economic stakes involved are huge. There is no branch of industry in which nanotechnologies will not be present in the near future.

The impact of nanotechnologies on the market has been the subject of many major studies. A parallel debate is taking place in the scientific communities as well as in the media, since the potential advantages are as important as the risks. Some examples of these are summarized in Table 10.2.

Many reports have been published recently about the nanotechnology issues regarding health, the environment and ethics. Even the insurance sector looked into this issue. Some of these warn us against rushing into these nanotechnologies, reminding us that the unpredictable nature of risks can generate significant loss.

Faced with the multitude of potential spin-offs from nanotechnologies, most current programmes are followed by significant research into their social, ethical and environmental impacts. This maybe illustrates that a lesson has been learnt from recent issues such as nuclear technology and GMO, where the risks were considered (respectively) after development or at the time of marketing. The sectors involved do not want the same pattern of rejection by the general public as GMO encountered, where many years of research and significant investment were lost. Therefore, above all else, it is a matter of caution in business.

Table 10.1   Crystalline and magnetic properties

Category	Number of products	Number in the USA	Number in Europe	Other countries
Cosmetics, beauty care, food etc.	58	28	UK: 8 France: 5	Korea: 7 Taiwan: 4 Australia: 3 Japan: 2 Israel: 1
Electronics	20	16		Japan: 3 Korea: 1
Air purifiers, deodorants, disinfectants etc.	30	16	UK: 1	Korea: 6 Taiwan: 3 Japan: 2
Textiles	25	18	Germany: 7	
Reinforcement of materials, surface treatment	57	41	Germany: 4 France: 2 Finland: 2 UK: 2 Sweden: 1	China: 1 Mexico: 1 Israel: 1 New Zealand: 1

Another reason to consider the social effect of nanotechnologies comes from the great diversity of sectors involved. For the general public this diversity is overwhelming and confusion about the differences in nanotechnologies develops. The application field of nanotechnologies is vast. According to their promoters, they will affect our daily lives, the world in which we are living and our bodies. Nanotechnologies bring up issues in many domains: health, the environment, the military and even bioengineering.

On top of this add the catastrophic scenario (or 'science-fiction' views) expressed by some promoters of nanotechnologies. The fears conveyed in these ideas get amplified by the philosophic, scientific and science-fiction literature. An example of this is the emblematic science-fiction book *Prey*, by Michael Crichton. In this story, nanomachines and hybrid bacteria create a swarm of nanorobots that act as a camera for reconnaissance and spying, with scientists losing control of their invention. All of these images together present a frightening (if unrealistic) image to the nonscientific community.

A report by a group of Canadian activists, ETC (Erosion Technology and Concentration), published in 2003, got a lot of media attention, as it revives fears of the dangers posed by nanotechnologies. This manifesto explains the different steps in the development of nanomaterials, their convergence with the living and the toxicity of nanoparticles, and even goes as far as to rename nanotechnologies 'atomtechnology'.

In the same year, Prince Charles also took up the issue by asking British scientists to analyse the risks of nanotechnologies to the environment and to society. In

Table 10.2   Expected benefits and potential risks of nanotechnologies

Field of application	Expected benefits	Potential risks
Technology of materials and surface	Miniaturization, increased performance, tailor-made materials, smart materials. Control of the biodegradation process, solar energy, material and energy saving.	Safety at work and human health in the use of nanoparticles and nanotubes.
Self-replicating material structure	Customized universal technique of molecular assembling. Development of programmable nanorobots to automatically fabricate molecular structures.	Uncontrolled proliferation of nanorobots capable of reproducing and scattering in the environment.
Biochemical implant in human organism	Smart implants capable of adapting to their environment and exchanging information with the outside world. Physical improvements for some people with disabilities.	Private-life protection. Extended social control. Human engineering. Discrimination between groups with various disabilities. Blurring of the boundary between the living and the artificial.
Medical nanotechnologies (DNA marker, nanopharmacy, biosensors	Preventive medicine. Early diagnostic. Individual specific pharmacy.	Omnipresent medicalization. Genetic coercion. Risk of social/medical exclusion.
Military technology	Transformation of human soldiers into invincible warriors (or cybersoldiers) through extensive use of implants. Remote control of humans on the battlefield.	Revival of the arms race. Risk of proliferation. Transformation of soldiers into exterminating robots.

particular they were to address the risk of 'grey goo', that is to say out-of-control, self-replicating nanorobots that consume entire ecosystems, resulting in global ecophagy: a situation comparable to the 'green goo', responsible for the advent of life on Earth. The extinction of humans and their habitat in favour of creatures stemming from industry (which humans have lost control of), obviously crystallizes all these fears.

These fears may make nanotechnologies specialists – scientists and engineers – smile. But they must avoid any arrogance; they must convince citizens that scientists, industrialists and politicians are genuinely taking their fears into account. The issues are concerned with important and diverse fields. Some of them (health, environment, military) raise concerns for the short term, while others are of concern in the long term.

10.1   Health and environmental issues

The economic implications of nanotechnologies are huge; however, some fears are naturally aroused concerning, in particular, their toxicity. In the media, the many categories of nanotechnology are grouped together, as if the risks were uniform across all of them. Some risks may exist, particularly in relation to the production of nanoparticles; however, these are not relevant to all nanotechnology applications.

It is worth noting, that, uniquely, these risks are being taken into account while the development of this sector is at a very early stage. Today, as this field is still in its infancy, studies on the effect to human life are few. However, it already seems that some precautions must (and will) be taken in certain cases.

Some fears have been raised concerning the properties of nanoparticles (which are used in applications such as high surface reactivity) and their potential effect on health given that they may be able to cross cell membranes. It is, however, recognized that most nanotechnologies do not present any particular risks. Only a limited number of materials present difficult problems, for instance the toxicity of some nanomaterials in fish brains. However, what is most feared by scientists is that effects like those that occurred with asbestos may exist. For instance, carbon nanotubes are extremely thin 'carbon fibres'; their properties differ from asbestos but will need to be fully analysed before any marketing of them can take place. Many of the applications for carbon nanotubes consist of them in dispersion in a solid matrix; however, it is not yet clear how easy their dispersal in air may be.

Not only carbon nanotubes, but also many other nanoparticles are, and will continue to be, inserted in solid matrices, in gels, cosmetics, paints, etc. In the case of cosmetics, for instance, all the effects related to the diffusion of nanoparticles through the skin or through scars are not known yet. It is acknowledged that some nanoparticles can be more toxic than their larger equivalent. These will need to be identified.

The way people will be exposed to dangerous nanomaterials must be determined. Most of them will be in factories or in enclosed places. Appropriate safety standards will have to be set for workers in these factories. Important issues exist concerning these safety standards for nanomaterials, regarding their fabrication, their transport, even their destruction (deliberate or, as in the case of fire, accidental). Today, two important authorities are contesting the pre-eminence of these safety standards: the American Society for Testing and Materials (ASTM) and the working party WG166 within the European Committee for Standardization (CEN, for Comité Européen de Normalisation).

Faced with this justifiable fear, the authorities have coordinated studies on the possible impact on health. Even if the budgets seem small (3 to 5 per cent of the investment), they are not negligible, and show the new state of mind in the relations between science, technology and society.

A commonly held idea in the media regarding nanoparticles is that the term refers to 'very tiny' and consequently this implies 'dangerous'. This generalization is obviously excessive. There are many nanoparticles in the air that are not dangerous to the human health. However, some studies must be conducted to evaluate the real risks of nanotechnologies. This must be done in cooperation with the manufacturers involved and those who work in the applied fields of biomedicine and pharmaceutics. This is recommended by, among others, Britain's Royal Society.

Environmental issues have many features in common with health, since the biological aspects have similarities with those relating to humans. Another additional issue is the effect of the dispersion of nanoparticles during their fabrication, in waste, and their effects on the environment. This is a very serious matter, which industrialists continue to monitor carefully.

Beyond the health and environmental aspects, some potential risks also require significant investment, particularly in the domain of safety. The health and environmental stakes will have to be understood before a real industrialization of nanotechnologies can occur.

10.2 Military interests

The military sector has always been interested in new scientific and technological developments. It is evident that the army takes an active interest in nanotechnologies. When the budgets dedicated to nanotechnologies are analysed, we can see that the military sector occupies a very important place. In the United States, in 2003, of the $774 million allocated to the National Nanotechnology Initiative (NNI), $243 million (i.e. 31.3 per cent) was allocated to the Department of Defense (DoD). In Europe, the figures are not as transparent, but the military contribution is still a significant proportion. What are the motives of the army? What are the risks related to these works?

Today, it is not predicted that nanotechnologies will specifically lead to new weapons. In the short term, nanotechnologies will be used in continuity with previous projects. The improvement of materials by the use of nanostructured materials will lead to lighter and stronger armour; it will cut down the maintenance costs and also improve the performances of kinetic-energy weapons (better penetration). In the electronics domain, the reduction of component size will allow sophisticated communication and detection systems, and allow them to be permanently operational thanks to improved energy sources.

Troops' health would be also improved by embedding sensors into their uniforms, giving (in real time) the state of their health in case chemical or bacterial weapons are used. In the medium term, the army will try to improve the performances of its combatants on the battlefield by using exoskeletons. The US Army expects that the deployment of troops with improved performances, far superior to those their enemies possess, will arouse fear, and consequently secure victory.

Nanotechnologies could also affect the development of biological, chemical and nuclear weapons. In the assessment of risks related to military research programmes, we will consider only those clearly linked to nanotechnologies. These risks are very diverse. They can be classified as *military*, *geostrategic* and *scientific*.

The purely military risks are related to the increasing complexity of the battlefield. The development and the deployment of increasingly sophisticated systems, working and communicating faster and faster, could make the battlefield unmanageable for humans.

At the geostrategic level, efforts in the developed countries to achieve new military applications could lead to undesirable situations. Therefore, other countries with important scientific resources will not remain inactive when faced with these developments. China and India have already increased their research budget in the nanotechnologies military domain. This consequently gives a justification for an increase in one's own military budget, and so the cycle continues to escalate.

Nanotechnologies also have the potential to put current treaties in jeopardy. Coupled with biotechnologies, nanotechnologies could lead to new techniques and new weapons that are not explicitly taken into account by current agreements.

Finally, when addressing scientific risks, these are not specific to nanotechnologies. The fact that scientific research is often carried out under military auspices often prevents the dissemination of information. Furthermore, the risks for health and environment are not negligible. In the military framework, the assessment of risks is not made public. As financial resources are limited, what is allocated to the army is necessarily withdrawn from civilian applications.

10.3   Media and ethical considerations

One of the biggest fears of scientists involved in nanotechnologies is that the excessive media coverage of certain aspects would lead, in the short or medium term, to antiscientific sentiment among the public. This fear is largely due to a controversy based on many misapprehensions, and is amplified by the poor scientific value of the information relayed by scientific or science-fiction literature to the general population.

In 1986, the year when Gerd Binnig and Heinrich Rohrer were awarded the Nobel Prize for having developed the scanning tunnelling microscope, Eric Drexler, a researcher at MIT, published one of the major books on nanotechnologies entitled *Engines of Creation*. This hypothesized that the development of sciences and technologies in the twentieth century would allow us to theorize and manipulate matter at the level of atoms and molecules. According to Drexler, it will be possible to reach another stage: to use living things at the nanometre scale. The role of the engineer would be to put together physics, chemistry and biology in the nanoworld to build nanomachines. These nanomachines, using the molecules and the properties of the living, would exponentially replicate in a medium where they would find the components for their bodies as well as components that would help for their assemblage, as living beings do.

The living becomes here a model and claim of feasibility. Moreover, since living things are endowed with consciousness, Eric Drexler thinks that engineers will be able even to engineer 'spirit' from the self-reproductive nanorobots, to remodel the body of man, to offer him support structures other than proteins in order that we might spread beyond the Earth. Biotechnologies bring – together with physics, chemistry and the information technologies necessary for programming nanorobots – the ability to take over living human beings. This breaks the established paradigm: our perception of nature and the body, our perception of synthetic and natural, our perception of prostheses and organs.

Even though most scientists do not believe in Drexler's ideas, some are even more ambitious. They see in the convergence of nanotechnologies, biotechnology, sciences, IT and cognitive technologies, a major programme known under the initialism NBIC.

10.4   NBIC

The convergence of these four major subjects (*n*anotechnologies, *b*iotechnologies, *i*nformation technology and *c*ognitive sciences) is an ambitious programme, which reintroduces the ancient ideal of philosophers of reuniting the sciences. According to Mihail Roc, one of the leading NBIC proponents,

> Half a thousand years ago, the Renaissance leaders mastered many domains at the same time. Today, on the other hand, the specialization has separated arts from engineering, and nobody can only master a small piece of the human creativity. Sciences have reached a stage where they must reunite in order to progress rapidly. The convergence of sciences can initiate a new Renaissance, incorporating a holistic conception of technology based on transforming tools, on mathematics of complex systems, and on a causal analysis of the unified physics from the nano-scale to the planetary scale.

It is also a matter of speeding up the process of cross-fertilization in order to obtain a synergistic effect that will produce something new, something more than the sum of its parts. Many various advantages for humanity are then considered. Let us look at some of the potential benefits that their promoters think possible.

- There would be the development of implants that would hybridize the body with machines in order to extend natural human senses and performances into an increased reality – an increase in sensitivity to the environment by an extended exchange of information with the brain, but also an increase in mnemic or calculus performances by the addition of logic units.
- Nanorobots could become our manufacturers, our depollutants, our energy providers, our remedies. They would hybridize with our ecosystems to give a nature that is attractive to humans, without diseases, one that is less hostile, serving our material wellbeing and desires.
- According to transhumanism, the human species has not yet reached its final stage; and it is the role of this technology, which is the natural extension of evolution, to give the body all the necessary malleability to evolve, in order to live in other places, to use other chemistry – all that is currently beyond even our extended realities.

This cannot be achieved without any risks. Some people dread a risk of toxicity for our bodies and habitats; they dread that ecosystems, hybridized by nanorobots, might develop beyond any control and that nanorobots could build an environment appropriate to their own proliferation but at humans' expense. This scenario is known as 'grey goo', following the example of the 'green goo' – oxygenation of our atmosphere consecutive to the development of plants. The grey goo has been popularized through Michael Crichton's book, *Prey*.

According to Bernadette Bensaude-Vincent, the risks generated by the NBIC programme are many, and are far beyond health and environmental problems. These risks are as follows.

- **To human liberty**: Nanotechnologies are invisible technologies. They allow and encourage the production of tiny and cheap sensors that can be implanted anywhere in order to watch or spy on people without their knowing. They take part in the process that puts the world under computer control. The implantation of chips under the skin is similar to the wide spread of biometry techniques and DNA tests or the radio-frequency identification device (RFID). Without necessarily believing that the initial intentions were such, one can fear a progressive extinction of our fundamental liberties.
- **To human integrity and dignity**: The coupling sensor-actuator makes possible manipulation of the will and behaviour by some implants.
- **To our private lives**: Nanorobots are likely to allow intrusion into private lives. Nanotechnologies abolish the notion of the private area and even the notion of actual locality.
- **To economic and political relations**: The risk of technology's spreading should not be underestimated, since nanotechnologies can be developed without expensive or sophisticated equipment.
- **To world peace**: The autonomous fighting devices and the spreading of toxic nanodusts make deterrent policies and treaties against nuclear proliferation ineffective.

All these risks and uncertainties obviously require the establishment of a 'safety-first' policy. However some doubts exist regarding the feasibility of the NBIC programme. In fact, in order for the NBIC to succeed, many conditions must be fulfilled simultaneously.

- To gather, around one single programme, people as diverse as physicists, mathematicians, chemists, biologists, engineers, computer scientists, specialists in communication sciences and cognitive sciences. Given that there is a diversity of mindsets and training across these fields, this programme would be a first and a great human challenge.
- To be successful in using a common language.
- To standardize actions that are very different.
- To understand and use the functioning of life and intelligence. Today, we do not really know how life evolved from the molecular state. Some conclude that, since living matter is composed of atoms and molecules, it should be possible to regenerate the living, and even intelligence, from engineering atoms. The 'spirit' would be only a combination of neurons communicating and exchanging information in bit form.
- To understand the complexity of life.

In these days, when the interest of youngsters in scientific disciplines is decreasing, while, on the other hand, the number of research subjects and new and promising applications increases exponentially, one might doubt that there will be enough people interested in such a big ambitious programme as the NBIC.

10.5 Education issues

However it may be, media coverage of the NBIC programme has been launched. Scientists can argue the fictitious nature of information reported by the media, and express concern that this media coverage may provoke unfounded fears, and sometimes disappointment due to lack of achievement. The media image of science would decline and the development of nanotechnologies would be uncertain due to hostility towards nanotechnologies.

As for the debate, it is desirable that it start very soon, before or at the same time as any new industrial venture begins. Such a venture would contribute to an increase in knowledge, although sometimes with disappointments. Like all behaviour, it would need a code of ethics, but this code should not be seen as a research restriction, more an edifying principle of behaviour, so that this venture might concentrate on the benefits. Otherwise, a futile or even dangerous conflict of interests would come about, because we would not be prepared to manage responsibly all the new choices we would face.

Media coverage could arouse fears, which would therefore become obstacles – a far cry from Eric Drexler's utopianism. Risks, inevitably, have to be managed. This risk management requires the appropriate training of scientists and citizens.

This issue of training is often neglected in the scientific sector. Between the researcher, the engineer and the citizen, there is often an important missing link we might call the 'translator' or the 'popularizer'. In the nanotechnology field, his or her role would not be simple. Nanotechnologies are based on concepts that are far from being intuitive and therefore easily understandable by the general public.

Another neglected training issue is that of future grey matter. In order to develop all these applications and those that come from all the other human activities (e.g. physics, chemistry, biology, medicine), scientists, engineers, technicians and biologists will be required. They will also be required in many other sectors, old and new, such as energy, transport, housing and nuclear fusion and fission. However, as mentioned earlier, the number of young scientists and engineers is decreasing in the developed world. Unless we have an aggressive policy of training and recruitment, one might well wonder what can actually be achieved, and at what pace. Moreover, as it is true that expanding countries such as China and India have an understandable interest in nanotechnologies, and these countries have enough grey matter to succeed, will Europe move on quickly enough?

Appendix 1
Electron microscopy

Only transmission electronic microscopy (TEM) will be discussed here. This technique represents a good choice for a tool for the study of nanomaterials, since it allows direct imaging of the samples. By observation of transversal sections, the images can be obtained both at the surface and inside the sample. Structural characteristics, such as shape, size and distribution of nanometric precipitates, can in this way be obtained. Furthermore, the advent of high-resolution techniques has allowed us to examine samples down to the atomic scale, and consequently to have a precise idea of their nanostructure.

Unlike the photons used in optical microscopy, electrons have a very small wavelength λ (which depends on the applied energy $E = hc/\lambda$). The energies usually used are in the range of 100–300 keV, which gives a theoretical optimal resolution of 0.2 nm, a value comparable to the characteristic interatomic distances of the matter. Therefore, this technique is perfectly adapted to the analysis of nanomaterials. However, observation is possible only in the case of very thin samples (<120 nm), which are transparent to the electronic beam. Therefore, thinning of the samples becomes necessary, the choice of which technique to use depending on the properties of the sample to analyse.

This imaging technique can be used for all kinds of materials, from metals to ceramics to polymers and semiconductors. It can be used, for instance, in the chemistry of carbon (nanotubes etc.), in polymer–silicate composites (see Chapter 8) and in silicon electronics. Biological materials (cells, DNA) can also be observed assuming that a cryogenic device is used to 'freeze' the material.

Figure A1.1 shows some cavities created by implantation of high doses of helium in silicon. This kind of cavity is the basis of the gettering technique, which involves the purification of the silicon by trapping the impurities present in the material.

Electronic diffraction techniques are also a very important tool for nanomaterials since the orientation of the selected grains can be deduced from the images: growing direction in the case of a thin film, particular orientation relations (known as *epitaxy*) between two grains of different crystallographic structures or different chemical composition. Figure A1.2 shows a high-resolution micrograph of a thin film of iron and the iron nanograins.

The associated electronic diffraction shows the polycrystalline state of the film. Each ring corresponds to a particular diffraction plane.

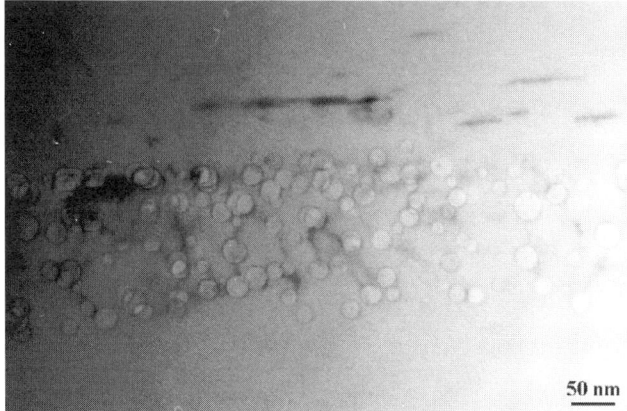

Figure A1.1   Band of cavities made by implantation of helium in silicon (500 °C, 5 × 1 016 cm⁻², 50 keV), which was then subjected to an annealing step (30 min at 800 °C) [Source: Laboratoire de Metallurgie Physique de Poitiers, M.-L. David, M.-F. Barbot]

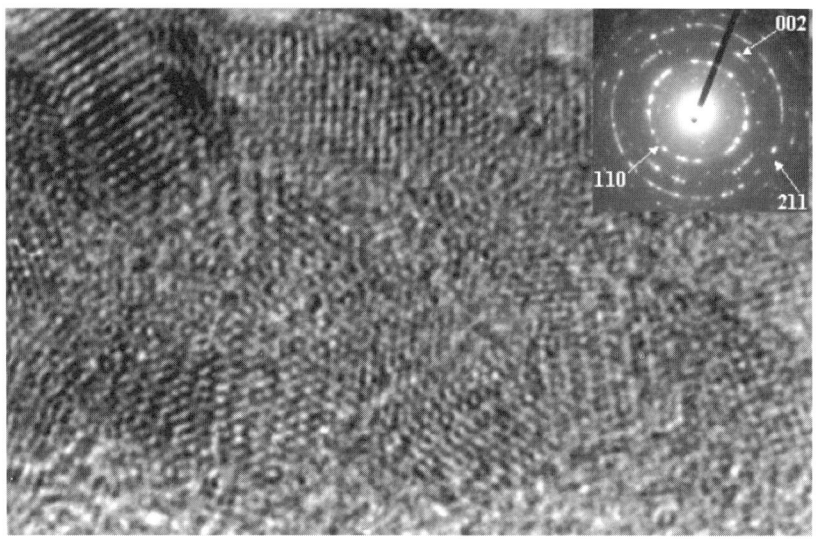

Figure A1.2   High-resolution micrograph of iron grains and corresponding electronic diffraction [Source: F. Monteverde, A. Michel, J.P. Eymery (private)]

Another advantage of electronic microscopy for the study of nanomaterials is the possibility of identifying precipitates of nanometric size by analysing their chemical composition, both qualitatively and quantitatively. Consequently, the electronic microscope allows focusing on grains of unknown nature, cartographies of a specific element on a predefined zone, or even the assessment of the stoichiometry of a specific zone.

To conclude, electronic microscopy techniques seem to have revolutionized and allowed the nanotechnologies domain to develop by their capacity to show the link between fabrication, structure and properties of nanomaterials, and all that down to the atomic scale.

Appendix 2
X photoemission spectroscopy (XPS) and secondary ions mass spectroscopy (ToF SIMS)

Physical effects that are important at our scale, such as gravity, are negligible at the micrometre scale and at the scale at which nanomaterials work. The molecular-attraction forces and the surface effects are predominant. For this reason, it is important to know the chemical nature of nanomaterials' surfaces perfectly. Therefore, surface-analysing techniques such as X photoemission spectroscopy (XPS) and time-of-flight secondary ions mass spectroscopy (ToF-SIMS) play a vital part and can be an efficient tool for the characterization of nanomaterials' surfaces. These two techniques can be used on a great variety of materials: metals, oxides, nitrates, carbides, organic compounds, etc., being either in bulk form as a thin film or in monolayer form.

A2.1   The XPS technique

This technique is based on the fact that each element existing in nature is composed of a nucleus surrounded by electrons with its own specific bond energy (E_{bond}). Moreover, according to the chemical environment of these elements in matter, some variations in the bond energy occur. When a material surface is subjected to the radiation of an RX monochromatic beam of energy hv (Al $K\alpha$ or Mg $K\alpha$), the energy of these incident photons is completely transmitted to the atoms making up the matter. Then these atoms are in an excited state and will emit some electrons, known as *photoelectrons*, which have a specific kinetic energy ($E_{kinetic}$). According to the principle of the conservation of energy, there is a direct relationship between this kinetic energy and the radiation energy:

$$E_{kinetic} = hv - E_{bond}$$

The XPS technique is based on the analysis of the kinetic energy emitted according to the process described earlier. The electrons emitted by the sample come from a very shallow depth, of the order of a few nanometres ($\approx$6 nm). From the detected peaks of photoelectrons, it is possible to determine the nature and the chemical state of the elements (Figure A2.1) present on the material surface and to determine the atomic percentage of these various elements. Their concentration is correlated to the intensity of the photoelectron peaks.

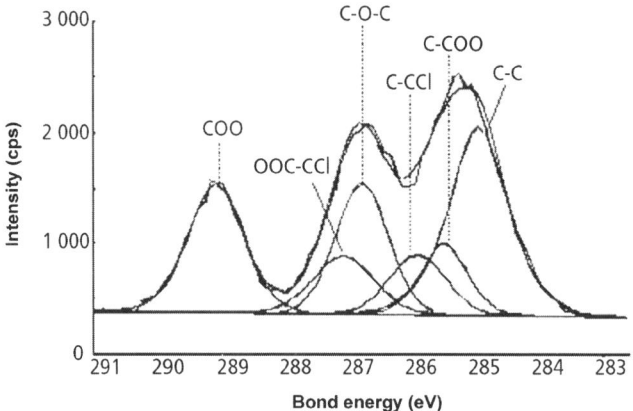

Figure A2.1   C 1s signal, characteristic of a thin film of poly(2-chloropropionate ethyl acrylate) electrochemically grafted on a steel surface [Source: M. Claes, S. Voccia, C. Detrembleur et al., submitted for publication in Macromolecule)

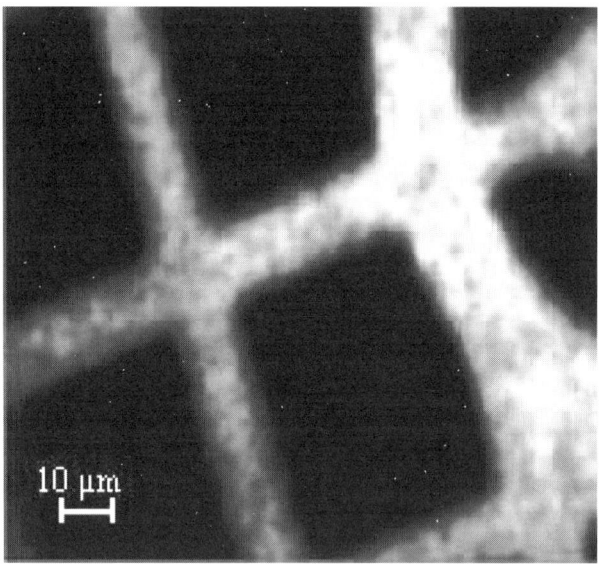

Figure A2.2   Distribution of copper at the surface of a microscope grid

It is also possible to visualise the distribution of compounds present at the material surface (Figure A2.2) by performing chemical cartography. Finally, by combining cycles of ion cleansing and XPS analysis, it is possible to obtain concentration profiles (Figure A2.3) on stacks of thin films.

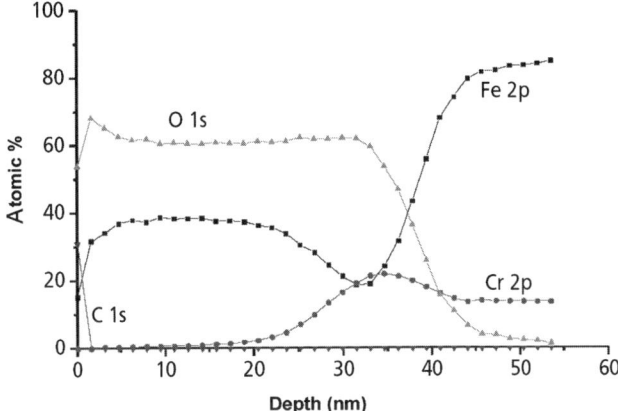

Figure A2.3 *Concentration profile of steel subjected to an oxygen plasma and to the implantation of O^+ and O_2^+ having an energy of 27 keV [Source: Lacoste, S. Bechu, Y. Arnal, J. Pelletier, R. Gouttebaron,* Surface and Coatings Technology, *156 (2002) pp. 225–8]*

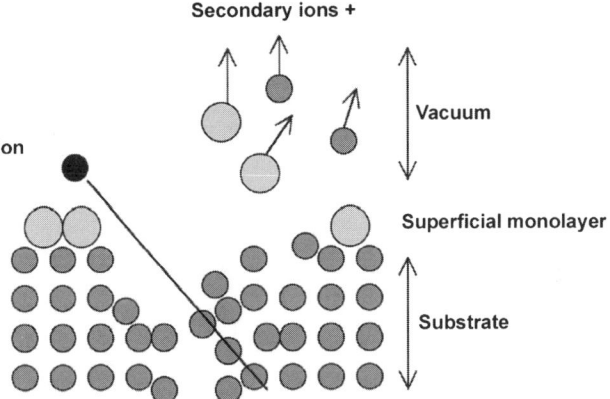

Figure A2.4 *Bombardment process*

A2.2 The ToF-SIMS technique

This technique is based on the irradiation of the surface of a material by a mono-energetic primary ion beam of a few dozen kV. A portion of these primary ions is reflected on the surface by a process of both elastic and non-elastic diffusion. The other ions penetrate into the material, there giving their energy to the atoms of the lattice by a series of collisions, and are finally implanted. Due to these collisions, the atoms of the lattice are displaced and the atoms of the superficial layers, receiving an impulse directed outside the target with an energy superior to the bond energy, are ejected by a bombardment process (see Figure A2.4). The bombarded material

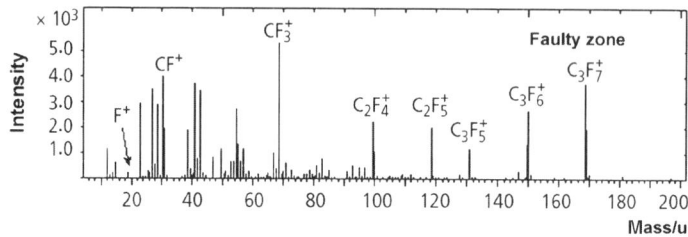

Figure A2.5   SIMS spectrum recorded at the surface of metal bodywork, at the centre of a defect, identification of the polluting agent as being a lubricant of the fabrication line [Source: Ion-tof GMBH]

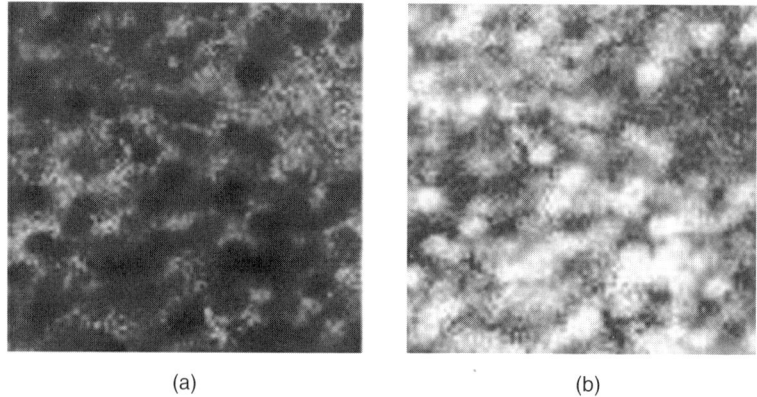

| (a) | (b) |

Figure A2.6   Nickel particles inside a polyethylene (PET) matrix: (a) distribution of the $C_2H_4^+$ ions, secondary ions specific to the PET matrix; (b) distribution of the N^+ ions [Source: M. Ferring, R. Vegis, D. Cornelissen]

is composed of different species: atoms and atom clusters present in the neutral, excited and ionized state (i.e. secondary ions, which represent a fraction of the order of 10^{-1} to 10^{-6} of the bombarded species). Furthermore, the bombarded material also contains electrons.

The ToF-SIMS analysis is based on the detection of the ejected secondary ions, by a time of flight mass spectrometer. Depending on the time required by the various secondary ions to cross the spectrometer and arrive at the detector, it is possible to determine precisely the mass of these ions. These detected secondary ions are atomic and molecular (in the case of the analysis of organic materials, clusters containing dozens, even hundreds, of atoms are detected), which gives information about the nature and the chemical structure of the surface compounds (see Figure A2.5) with great detection sensitivity (of the order of ppm). However, it is worth mentioning that the chemical environment of an element has a great influence on the yield of

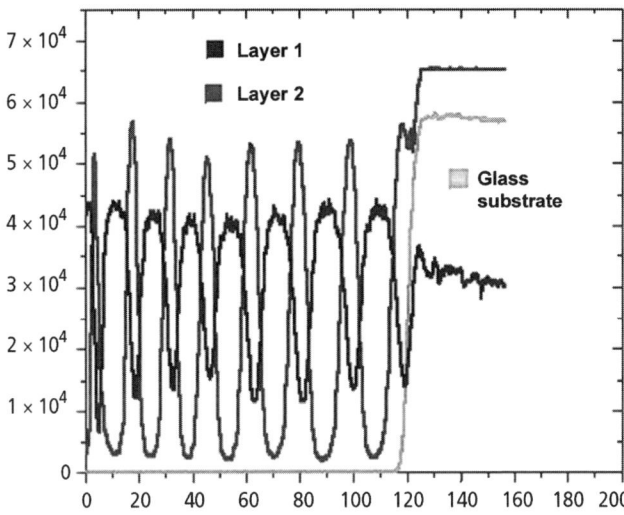

Figure A2.7   SIMS profile of a multilayer device deposited on top of a glass substrate surface [Source: Ion-tof GMBH]

the bombardment of these specific ions (matrix effects). Therefore, this analysis technique is qualitative rather than quantitative.

After identification of the ions, it is possible to visualize the distribution of the compounds by a chemical cartography of the materials' surface with a lateral resolution of the order of micrometres. In order to achieve such an analysis, the mass of the different compounds to be visualized must be known. Then the intensity of the peak of the various compounds is stored pixel by pixel, while the primary ion beams (reduced to a spot of a small diameter, a microbeam of 0.15 μm) scan the surface line by line. The secondary optics for extraction and mass analysis are fixed. The image is rebuilt by synchronization of the detected secondary signal with the scanning of the primary beam (see Figure A2.6).

Finally, it is possible to make some analyses in depth by combining the Tof-SIMS analysis with some cycles of bombardment of the material by a second incident ion beam having an ion density greater than 10^{16} ions/cm^2, which allows us to pulverize the matter rapidly (Figure A2.7).

Appendix 3
Imaging by nuclear magnetic resonance

The main application of the nuclear magnetic resonance (NMR) phenomenon is indisputably magnetic resonance imaging (MRI). This technique, devised in 1973 by P.C. Lauterbur to visualize a system of capillaries full of water, has been widely used for more than 15 years in the medical environment as a diagnosis tool. It allows us to have contrasted anatomic images from the NMR signal of the protons of human bodies. The contrast is imposed by the protons' density of the tissues, by the longitudinal (T_1) and transverse (T_2) relaxation time, and by the acquisition parameter of the image.

A3.1   Relaxation times

The relaxation times play an especially important role in the contrast between tissues. They characterize the return to thermodynamic equilibrium of a sample whose magnetization has been shifted away from its equilibrium value. This return is caused by fluctuations of the local magnetic field to which each proton is subjected. This shift influences the direction of the magnetization vector (aligned with the external static field at equilibrium) as well as its modulus. T_1, the longitudinal relaxation time, influences the return to equilibrium of the magnetization component parallel to the static field, while T_2, the transverse relaxation time, influences the disappearance of the component perpendicular to the static field. In general, T_1 is sensitive only to the relatively rapid fluctuations of the local field, while T_2 is determined by all fluctuations, rapid as well as slow. Contrast agents shorten these relaxation times, so they do not appear on the images, their role being only indirect – which justifies their names as contrast agents rather than contrast materials. They are classified in two different groups: positive agents (acting preferentially on T_1) and negative agents (acting preferentially on T_2). In fact, shortening the longitudinal relaxation time causes an increase in the initial magnetization from which the signal decreases, while shortening the transverse relaxation time causes a decrease of the observed signal, since to produce this signal it is the transverse component of the magnetization that is detected.

Compared with X-ray tomography, MRI has two important advantages: it gives a better contrast and the magnetic field (static and oscillating) used seems to be completely harmless, which is clearly not the case for X-rays.

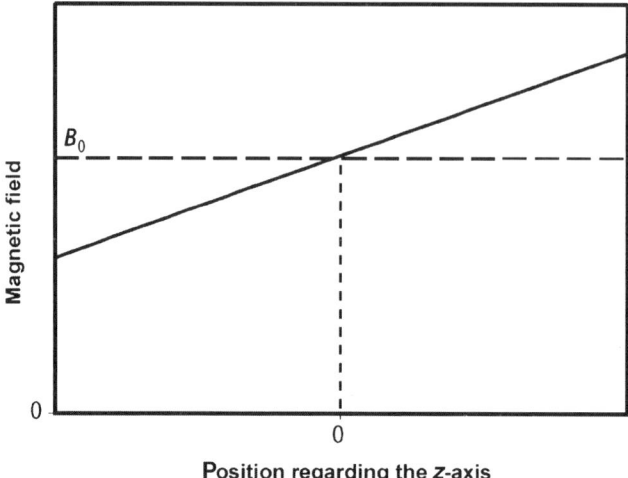

Figure A3.1 Magnetic field gradient along the z-axis

A3.2 Selection of a slice and voxel notion

In order to make an image of any part of the body, from head to foot, we must be able to choose the section of our choice. To this end, magnetic-field gradients are used. These gradients are added to the constant magnetic field B_0. Consequently, when a constant gradient along z is applied, the intensity of the magnetic field B to which the protons are subjected increases linearly with their position along z. Subsequently, the Larmor frequency ($\nu = \gamma B/2\pi$) of the protons of the patient's foot is different from that of the protons of his head. The Larmor frequency is the resonance frequency of the protons. They are excited only if an electromagnetic wave of the correct frequency ν is used. Then, it is the localization of the resonance that allows us to overcome the issue (reputed to be insurmountable beforehand) of the poor resolution due to the use of radiofrequency electromagnetic waves (wavelength of the order of a dozen metres).

A radiofrequency wave of frequency $[\upsilon - \frac{\delta}{2}; \upsilon + \frac{\delta}{2}]$ will then have an effect only on protons whose Larmor frequency is situated in this range, i.e. protons situated in a slice of finite thickness perpendicular to z (axial section). After the radiofrequency excitation impulse, only the magnetic moment of the protons of this slice is excited and contributes to the signal (Figure A3.2). To select another slice, modifying the frequency of the radiofrequency impulse is enough.

If a gradient along y is used, a slice corresponding to a sagittal section is selected. It is worth mentioning that the impulses used in imaging have a shape elaborated in order to obtain an optimal definition of the excited section.

In order to record an image of the selected slice, it is necessary to distinguish the signal coming from the different volume elements (voxels) composing the slice. In fact, it is the differences in signal intensities between near voxels that are at the

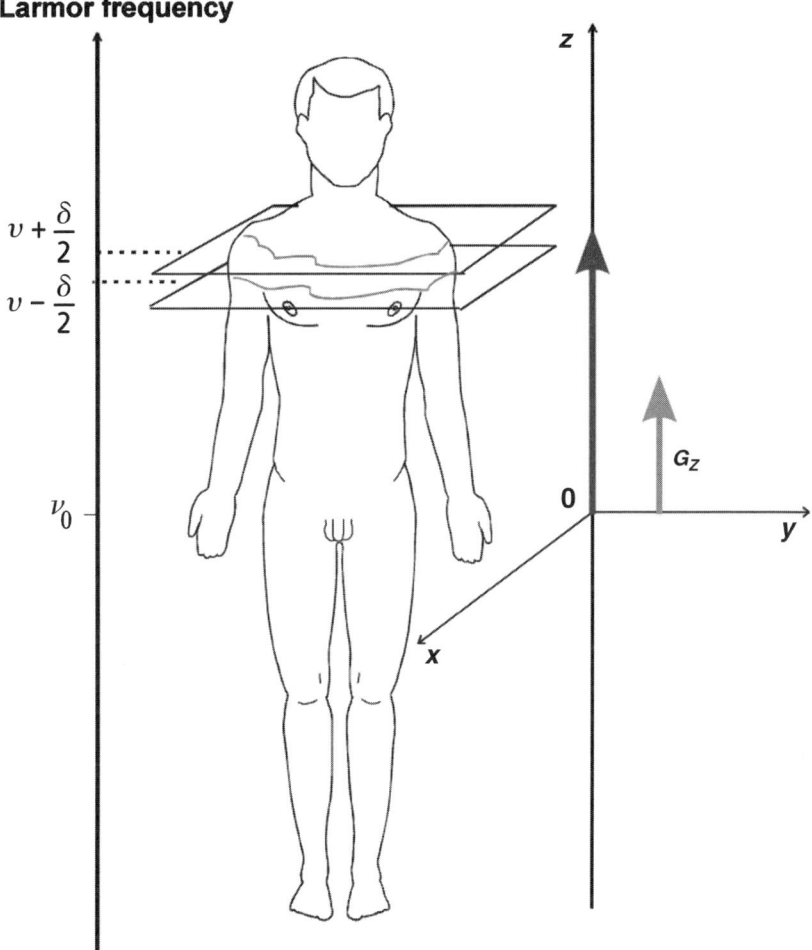

Larmor frequency

$v + \dfrac{\delta}{2}$

$v - \dfrac{\delta}{2}$

v_0

Figure A3.2   Principle of the slice selection

root of the contrast in the image. To this end, some magnetic-field gradients along the x-axis and the y-axis are applied, at specific times of the measurement sequence. Then, a computerized data processing of the detected signal (a two-dimensional Fourier transform) gives a contrasted image of the selected slice.

Bibliography

General

Bhushan, B. (ed.), *Springer Handbook of Nanotechnology* (Springer-Verlag, Berlin, 2004)

Corriu, R., Nozières, P., and Weisbuch, C., *Nanosciences et Nanotechnologies* (Tec&Doc, Paris, 2004)

Lahmani, M., Dupas, C., and Houdy, P., *Les nanosciences. Nanotechnologies et nanophysique* (Belin, Paris, 2004)

Chapter 1

Elwenspoek, M., and Wiegerink, R., *Mechanical Microsensors* (Springer-Verlag, Berlin, 2001)

Pautrat, J.-L., *Demain le nanomonde* (Fayard, Paris, 2002)

Sargent, T., *Bienvenue dans le nanomonde* (Dunod, Paris, 2006)

Timp, G. (ed.), *Nanotechnology* (Springer-Verlag, New York, 1999)

Chapter 2

Desjonquères, M.C., and Spanjaard, D., *Concepts in Surface Physics* (Springer-Verlag, Berlin, 1993)

Kendall, K., 'Adhesion: Molecules and Mechanics', *Science*, 1994;**263**:1720

Lucas, A.A., Moreau, F., and Lambin, Ph., 'Optical simulations of electron diffraction by carbon nanotubes' *Revs. Mod. Phys.*, 2002;**74**:1

Moriarty, P., 'Nanostructured materials', *Repts. Progr. Phys.*, 2001;**64**:297

Sugano, S., *Microcluster Physics* (Springer-Verlag, Berlin, 1991)

Timp, G. (ed.), *Nanotechnology* (Springer-Verlag, Berlin, 2000)

Yacaman M.J., *et al., J. Vac. Sci. Technol.*, 2001;**B 19**:1091

Chapter 3

Gaponenko, S.V., *Optical Properties of Semiconductor Nanocrystals* (Cambridge, Cambridge University Press, 1998)

Jacak, L., Hawrylak, P., and Wojs, A., *Quantum Dots* (Springer-Verlag, Berlin, 1998)

Kittel, C., *Physique de l'État Solide*, 8th edn (Dunod, Paris, 2007)

Timp, G. (ed.), *Nanotechnology* (Springer-Verlag, New York, 1999)

Chapter 4

Collins, P.G., and Avouris, P., 'Nanotubes for Electronics', *Scientific American*, December 2000, p. 38

Jacoby, M., 'Nanoscale Electronics', *Chemical & Engineering News*, September 2002, p. 38

Ratner, M.A., 'Introducing Molecular Electronics', *Materials Today*, February 2000, p. 20

Reed, M.A., and Tour, J.M., 'Computing with Molecules', *Scientific American*, June 2000, p. 68

Chapter 7

Brittain, S., Paul, K., Zhao, X.M., and Whitesides, G.M., 'Soft lithography and microfabrication', *Physics World*, 1998;**11**:31–6

IBM project *Millipede* [online]. Available from http://domino.research.ibm.com/comm/pr.nsf/pages/rsc.millipede.html

IBM project *Scanning Tunneling Microscopy Image Gallery* [online]. San Jose, California: Almaden Research Center Visualization Lab. Available from http://www.almaden.ibm.com/vis/stm/gallery.html

Vettiger, P., Despont, M., Drechsler, U., *et al.*, 'The "Millipede" – Nanotechnology Entering Data Storage', *IEEE Trans. Nanotechnol.*, 2002;**1**(1):39–55

Xia, Y., and Whitesides, G.M., 'Soft lithography', *Annu. Rev. Mater. Sci.*, 1998, pp. 153–84

Chapter 8

Alexandre, M., and Dubois, P., 'Polymer-layered silicate nanocomposites: preparation, properties and uses of a new class of materials', *Materials Science and Engineering Reports* 28 (Elsevier Science, New York, 2000)

Janot, C., and Ilschner, B., *Matériaux émergents*, Traité des matériaux series, Vol. 19 (Presses polytechniques et universitaires romandes, Lausanne, 2001)

Pinnavaia, T.J., and Beall, G.W., *Polymer-clay Nanocomposites* (John Wiley & Sons, Chichester, 2000)

Van Damme, H., *Nanomatériaux*, Série Arago No. 27 (Observatoire français des techniques avancées, Paris, 2001)

Chapter 9

Blundell, S., *Magnetism in Condensed Matter* (Oxford University Press, 2001)

De Cuyper, M., and Joniau, J., 'Magnetoliposomes: Formation and structural Characterization', *Eur. Biophys. J.*, 1988;**15**:311–19

Frankel, R.B., and Blakemore, R.P., 'Magnetite and magnetotaxis in bacteria', *Bioelectromagnetics*, 1989;**10**:223–37

Gelman, N., Gorell, J.M., Barker, P.B., *et al.*, 'MR imaging of human brain at 3.0 T: preliminary report on transverse relaxation rates and relation to estimated iron content', *Radiology*, 1999;**210**:759–67

Halbreich, A., Roger, J., Pons, J.N., *et al.*, 'Biomedical applications of maghemite ferrofluid', *Biochimie*, 1998;**80**:379–90

Jordan, A., Scholz, R., Maier-Hauff, K., *et al.*, 'Presentation of a new magnetic field therapy system for the treatment of human solid tumors with magnetic fluid hyperthermia', *Journal of Magnetism and Magnetic Materials*, 2001;**225**:118–26

Jordan, A., Wust, P., Fahling, H., *et al.*, 'Inductive heating of ferrimagnetic particles and magnetic fluids: physical evaluation of their potential for hyperthermia', *Int. J. Hyperthermia*, 1993;**9**:51–68

Kirschvink, J.L., Walker, M.M., and Diebel, C.E. 'Magnetite-based magnetoreception', *Curr. Opin. in Neurobiol.*, 2001;**11**:462–67

Merbach, A.-E., and Toth, E., *The Chemistry of Contrast Agents in Medical Magnetic Resonance Imaging* (John Wiley & Sons, Chichester, 2001)

Molday, R.S., *Magnetic Iron-Dextran Microspheres*: US Patent, International Patent Number 4 452 773, 1984

Nakatsuka, K., 'Trends of magnetic fluid applications in Japan', *J. Magn. Magn. Mater.*, 1993;**122**:387–94

Néel, L., 'Théorie des propriétés magnétiques des grains fins antiferromagnétiques', in de Witt C. (ed.), Conférence à l'École de Physique Théorique, Les Houches, CNRS, Paris, 1961

Rosensweig, R.E., 'Heating magnetic fluid with alternating magnetic field', *J. Magn. Magn. Mater.*, 2002;**252**:370–74

Webb, J., Macey, D.J., Chua-Anusorm, W., *et al.*, 'Iron biominerals in medicine and the environment', *Biochim. Biophys. Acta*, 1999;**190–192**:1199–215

Chapter 10

Bensaude-Vincent, B., *Se libérer de la matière? Fantasmes autour des nouvelles technologies* (INRA, Paris, 2004)

Bensaude-Vincent, B., 'Two Cultures of Nanotechnology?', *HYLEInt. J. Phil. Chem.*, 2004;**10**:67–84

Brooker, R., and Boysen, E., *Nanotechnology for Dummies* (Wiley, Indianapolis, 2005)

Drexler, E., *Engines of Creation. The coming era of nanotechnology* (Anchor Books, New York, 1986)

Appendix 3

Vlaardingerbroek, M.T., and Den Boer, J.A., *Magnetic Resonance Imaging* (Springer Verlag, Berlin Heidelberg, 1996)

Index